Innovative Approaches in Earl Childhood Mathematics

The chapters in this book investigate and reflect on many of the issues and challenges raised by the current trends and tensions in early childhood mathematics education. They emanate from seven countries – Australia, Northern Ireland, Norway, Portugal, Spain, Sweden, and Switzerland.

Ever since Fröbel invented the kindergarten, mathematics has been a part of early childhood pedagogy. Mathematics is an important part of children's daily life, which helps them to understand the world around them. Nowadays, early childhood mathematics is in the international spotlight. Partly this is the result of myriad studies that seem to show that early childhood mathematics achievement is a strong predictor of success or otherwise in future school mathematics, other school subjects, and life itself. Another influence on early childhood mathematics education is the advent of the political and advocacy juggernaut known as STEM (Science, Technology, Engineering, and Mathematics). Early childhood mathematics education is important for children's present and future learning.

This book provides a strong collection of current research for the consideration of all in the early childhood education field. It was originally published as a special issue of the *European Early Childhood Education Research Journal*.

Oliver Thiel is an Associate Professor at Queen Maud University College in Trondheim, Norway. He has 20 years of teaching and research experience in the field of early childhood mathematics education in Germany and Norway. He is convenor of the EECERA Special Interest Group: Mathematics Birth to Eight Years.

Bob Perry is recently retired after 45 years in higher education. He is Emeritus Professor in the School of Education at Charles Sturt University in Albury-Wodonga, Australia, and Director of Peridot Education Pty Ltd. He has been awarded an Honorary Doctorate from Mälardalen University, Sweden, in recognition of his body of research.

EECERA Collection of Research in Early Childhood Education

Written in association with the European Early Childhood Education Research Association (EECERA), titles in this series will reflect the latest developments and most current research in early childhood education on a global level. Feeding into and supporting the further development of the discipline as an exciting and urgent field of research and high academic endeavour, the series carries a particular focus on knowledge creation and reflection, which has huge relevance and topicality for those at the front line of decision making and professional practice in early years services.

Titles in this series:

Early Childhood Education and Change in Diverse Cultural Contexts
Edited by Chris Pascal, Tony Bertram and Marika Veisson

Early Childhood Care and Education at the Margins
African Perspectives on Birth to Three
Edited by Hasina Banu Ebrahim, Auma Okwany and Oumar Barry

Perspectives from Young Children on the Margins
Edited by Jane Murray and Colette Gray

Working with Parents and Families in Early Childhood Education
Edited by Ute Ward and Bob Perry

Innovative Approaches in Early Childhood Mathematics
Edited by Oliver Thiel and Bob Perry

For more information about this series, please visit:
https://www.routledge.com/EECERA-Collection-of-Research-in-Early-Childhood-Education/book-series/EECERARES

Innovative Approaches in Early Childhood Mathematics

Edited by
Oliver Thiel and Bob Perry

Routledge
Taylor & Francis Group

LONDON AND NEW YORK

First published 2020
by Routledge
2 Park Square, Milton Park, Abingdon, Oxon, OX14 4RN

and by Routledge
52 Vanderbilt Avenue, New York, NY 10017

Routledge is an imprint of the Taylor & Francis Group, an informa business

First issued in paperback 2021

Introduction, Chapters 2, 4–6, 8, 10 © 2020 EECERA
Chapter 1 © 2018 Camilla Björklund, Maria Magnusson and Hanna Palmér. Originally published as Open Access.
Chapter 3 © 2018 Gabriella Gejard and Helen Melander. Originally published as Open Access.
Chapter 7 © 2018 Hanna Palmér and Jorryt van Bommel. Originally published as Open Access.
Chapter 9 © 2018 Franziska Vogt, Bernhard Hauser, Rita Stebler, Karin Rechsteiner and Christa Urech. Originally published as Open Access.

British Library Cataloguing in Publication Data
A catalogue record for this book is available from the British Library

ISBN13: 978-0-367-35410-7 (hbk)
ISBN13: 978-1-03-208588-3 (pbk)

Typeset in Minion Pro
by Newgen Publishing UK

Publisher's Note
The publisher accepts responsibility for any inconsistencies that may have arisen during the conversion of this book from journal articles to book chapters, namely the inclusion of journal terminology.

Disclaimer
Every effort has been made to contact copyright holders for their permission to reprint material in this book. The publishers would be grateful to hear from any copyright holder who is not here acknowledged and will undertake to rectify any errors or omissions in future editions of this book.

Contents

Citation Information

The chapters in this book were originally published in the *European Early Childhood Education Research Journal*, volume 26, issue 4 (August 2018). When citing this material, please use the original page numbering for each article, as follows:

Introduction

Chapter 1

Chapter 2

Chapter 3

Chapter 4

Chapter 5

Mathematical pedagogical content knowledge in Early Childhood Education: tales from the 'great unknown' in teacher education in Portugal
Maria Pacheco Figueiredo, Helena Gomes and Cátia Rodrigues
European Early Childhood Education Research Journal, volume 26, issue 4 (August 2018) pp. 535–546

Chapter 6

The impact of the Promoting Early Number Talk project on the development of abstract representation in mathematics
Pamela Moffett and Patricia Eaton
European Early Childhood Education Research Journal, volume 26, issue 4 (August 2018) pp. 547–561

Chapter 7

The role of and connection between systematization and representation when young children work on a combinatorial task
Hanna Palmér and Jorryt van Bommel
European Early Childhood Education Research Journal, volume 26, issue 4 (August 2018) pp. 562–573

Chapter 8

What makes a task a problem in early childhood education?
Rafael Ramírez-Uclés, Elena Castro-Rodríguez, Juan Luis Piñeiro and Juan F. Ruiz-Hidalgo
European Early Childhood Education Research Journal, volume 26, issue 4 (August 2018) pp. 574–588

Chapter 9

Learning through play – pedagogy and learning outcomes in early childhood mathematics
Franziska Vogt, Bernhard Hauser, Rita Stebler, Karin Rechsteiner and Christa Urech
European Early Childhood Education Research Journal, volume 26, issue 4 (August 2018) pp. 589–603

Chapter 10

Using a bioecological framework to investigate an early childhood mathematics education intervention
Bob Perry and Sue Dockett
European Early Childhood Education Research Journal, volume 26, issue 4 (August 2018) pp. 604–617

For any permission-related enquiries please visit:
www.tandfonline.com/page/help/permissions

Notes on Contributors

Camilla Björklund is a Professor in the Department of Education, Communication and Learning at the University of Gothenburg, Sweden.

Elena Castro-Rodríguez works in the Department of Didactics of Mathematics at the University of Granada, Spain.

Christina Davidson is a Senior Lecturer in Literacy Education in the School of Education at Charles Sturt University, Australia.

Sue Dockett is an Emeritus Professor of Early Childhood Education in the School of Education at Charles Sturt University, Australia.

Patricia Eaton is Director of Teaching and Learning in the Department of Teacher Education at Stranmillis University College, UK.

Angela Fenton is a Senior Lecturer and Associate Head of the School of Education at Charles Sturt University, Australia.

Maria Pacheco Figueiredo is an Assistant Professor in the School of Education and CI&DETS at the Polytechnic Institute of Viseu, Portugal.

Gabriella Gejard is a Senior Lecturer in the Department of Education at Uppsala University, Sweden.

Helena Gomes works in the School of Education at the Polytechnic Institute of Viseu, Portugal.

Bernhard Hauser is a Lecturer and Scientific Assistant at the Institute for Research in Teaching and Learning at the University of Education St. Gallen, Switzerland.

Lars Jenssen is a Researcher and Science Coordinator in the Department of Primary Education at Humboldt University of Berlin, Germany.

Amy MacDonald is a Senior Lecturer in Early Childhood Mathematics Education in the School of Education at Charles Sturt University, Australia.

Maria Magnusson is a Senior Lecturer in the Department of Education and Teachers' Practice at Linnaeus University, Sweden.

Helen Melander is a Senior Lecturer in the Department of Education at Uppsala University, Sweden.

Pamela Moffett is a Senior Lecturer in Mathematics with Curriculum Studies in the Department of Teacher Education (Primary) at Stranmillis University College, UK.

Hanna Palmér is an Associate Professor in the Department of Mathematics at Linnaeus University, Sweden.

Bob Perry is Emeritus Professor in the School of Education at Charles Sturt University in Albury-Wodonga, Australia, and Director of Peridot Education Pty Ltd.

Juan Luis Piñeiro is a PhD student in the Department of Didactics of Mathematics at the University of Granada, Spain.

Rafael Ramírez-Uclés works in the Department of Didactics of Mathematics at the University of Granada, Spain.

Karin Rechsteiner is a Lecturer at the Institute for Research in Teaching and Learning at the University of Education St. Gallen, Switzerland.

Cátia Rodrigues works in the School of Education at the Polytechnic Institute of Viseu, Portugal.

Juan F. Ruiz-Hidalgo works in the Department of Didactics of Mathematics at the University of Granada, Spain.

Rita Stebler is a Research Associate at the Institute of Education Research at the University of Zurich, Switzerland.

Oliver Thiel is an Associate Professor at Queen Maud University College in Trondheim, Norway. He is convenor of the EECERA Special Interest Group: Mathematics Birth to Eight Years.

Christa Urech is a Lecturer at the Institute for Research in Teaching and Learning at the University of Education St. Gallen, Switzerland.

Jorryt van Bommel is an Assistant Professor in Mathematics Education in the Department of Mathematics and Computer Science at Karlstad University, Sweden.

Franziska Vogt is a Professor and Director of the Institute for Research in Teaching and Learning at the University of Education St. Gallen, Switzerland.

Introduction – Innovative approaches in early childhood mathematics

Oliver Thiel and Bob Perry

Congratulations to the European Early Childhood Education Research Association (EECERA) Special Interest Group on Mathematics Birth to Eight Years for the genesis of this special issue of the *European Early Childhood Education Research Journal* (EECERJ). The idea for a special issue about early childhood mathematics emerged first at the EECERA meeting in Barcelona in 2015 and solidified into a firm proposal in Dublin in 2016. It has been a long time coming but we are very proud of its final form. Thank you to the members of the SIG who have contributed, not only through writing but through their enthusiastic backing, and thank you to other contributors from outside the SIG. The result is an excellent collection of papers based on rigorous, sensitive and timely research on innovative approaches to early childhood mathematics.

Ever since Friedrich Fröbel (1862) invented the kindergarten, mathematics has been a part of early childhood pedagogy. Fröbel was aware of the educational potential in play and games and developed his 'Spielgaben' (German = play gifts, in English called Froebel Gifts) – toys that embody mathematical ideas such as symmetry, shape, and number (Fröbel and Lilley 1967; von Marenholtz-Bülow 1887). He knew that mathematics is an important part of every child's daily life which helps them to understand the world around them. In the twentieth century, working with mathematics in early childhood was mostly play based and rather implicit, and learning occurred incidentally. In Nordic and Central European countries that follow a social pedagogy tradition (Bennett and Taylor 2006), pre-school focused primarily on social skills and care rather than education (Hemmerling 2007). This has changed in recent decades, especially after the 'PISA-Shock' which led to an international systematisation in education, a global standards movement with a shift in policy focus from educational inputs to learning outcomes, and an increase in educational research and measurement (Gruber 2006).

Nowadays, early childhood mathematics is in the international spotlight. Partly this is the result of a myriad of studies that seem to show that early childhood mathematics achievement is a strong predictor of success or otherwise in future school mathematics, other school subjects and life itself (Duncan et al. 2007; Geary et al. 2013; Carmichael, MacDonald, and McFarland-Piazza 2014). As a result, across the globe, there is greater encouragement for early childhood professionals in both prior-to-school and school settings to engage with their children in mathematics learning, with one aim being to ensure that the children's standards of achievement are higher by the time they meet the first national or international assessment of their careers. As Peter Moss (2014) has noted one of the results of such 'encouragement' has been the 'schoolification' of prior-to-school education and moves away from play-based pedagogies – a tendency that many early childhood professionals meet with scepticism (Broström 2017).

Another influence on early childhood mathematics education, which is related to the standards-based arguments, is the advent of the political and advocacy juggernaut known as STEM (Science, Technology, Engineering, and Mathematics). There is a danger that, as a result of STEM advocacy, mathematics will be seen to be the 'servant' of science, technology, and engineering and that all mathematics will need to be drawn from these other disciplines or apply to them. However, mathematics, particularly mathematical thinking, has a nature and approach which demand respect in its own right (Hardy 1940; Devlin 2012). At the early childhood level, mathematics provides opportunities for investigation and discovery that are not limited to applications to 'real' life but also stimulate creative and innovative thinking in both young children and their educators (Shen and Edwards 2017). Mathematics must not become simply a servant of science but rather be an approach to thinking and reasoning for young children's present and future (Katz 2010).

The first decade of this century saw the development of two position statements on early childhood mathematics – one in the U.S.A. and one in Australia – which are still pertinent today. The first resulted from a joint project between the National Association for the Education of Young Children (NAEYC) and the National Council of Teachers of Mathematics (NCTM), two U.S.-based professional associations which did not have a history of working together and furnishes a strong position for early childhood mathematics education.

> The National Council of Teachers of Mathematics and the National Association for the Education of Young Children affirm that high-quality, challenging, and accessible mathematics education for three-to-six-year-old children is a vital foundation for future mathematics learning. In every early childhood setting, children should experience effective, research-based curriculum and teaching practices. Such high-quality practice in turn requires policies, organizational supports, and adequate resources that enable teachers to do this challenging and important work. (NAEYC & NCTM 2002/2010, 1)

In Australia, the equivalent professional associations – Early Childhood Australia (ECA) and the Australian Association of Mathematics Teachers (AAMT) took the following position:

> The Australian Association of Mathematics Teachers and Early Childhood Australia believe that all children in their early childhood years are capable of accessing powerful mathematical ideas that are both relevant to their current lives and form a critical foundation for their future mathematics and other learning. Children should be given the opportunity to access these ideas through high quality child-centred activities in their homes, communities, prior-to-school settings and schools. (ECA & AAMT 2006, 1)

We have chosen to include these statements in this editorial partly because they originate outside the European context of EECERA and can be compared with perhaps more familiar documents for many readers, but, more importantly, because they reiterate the importance of early childhood mathematics education for both the present and the future learning and advocate for the children, educators and approaches involved in this learning (see also Moss, Bruce, and Bobis 2016). The papers presented in this special issue of EECERJ investigate and reflect many of the issues and challenges raised by the position statements and the current trends and tensions in early childhood mathematics education. They provide a strong collection of current research for the consideration of all in the early childhood education field.

The 10 papers in this special issue emanate from seven countries – Switzerland, Sweden (three papers), Spain, Portugal, Norway, Northern Ireland, and Australia (two). The papers address many of the 'trending' topics in early childhood mathematics education and provide important insights to these topics.

The first three papers in this issue address various aspects of the important mathematical idea of 'mathematization', 'a term coined by the eminent Dutch mathematics educator, Hans Freudenthal, in the 1960s to signify the process of generating mathematical problems, concepts and ideas from a real world situation and using mathematics to attempt a solution to the problems so derived' (Perry and Dockett 2008, 81). Björklund, Magnusson, and Palmér (Sweden) use the framework of Developmental pedagogy (Samuelsson and Carlsson 2008) to relate this idea to the common early childhood pedagogical approach of play. They highlight the need for teachers to interact with children during their play in order to help them understand their worlds and to mathematize. The analysis reveals four different lines of action that teachers can use to stimulate children's learning based on their play experiences. Through such an analysis, they develop a way of considering play and learning not in dichotomous opposition to each other but in harmonious mathematization. From the other side of the world, MacDonald, Fenton, and Davidson (Australia) investigate the mathematics arising from an experience in which many children across the world engage – shopping. In particular, they consider what mathematics children and their families notice, explore, and talk about as they participate in family shopping experiences. All six of Bishop's (1988) cultural mathematical practices were noticed, suggesting that shopping may be one example of an experience which could be used by both families and early childhood professionals as a starting point for mathematizing play in the sense introduced by Björklund, Magnusson, and Palmér. The third paper in this group also emanates from Sweden and, like the first two, relies on the analysis of videorecorded data in its investigation of children's mathematising in their spatial play. Gejard and Melander use the notion of mathematizing as 'participation in mathematical discourse' (Sfard 2008, 128) to conduct a fine-grained study of the mathematical discourse when two pre-school children play with a magnetic construction toy. They emphasise that their findings, while preliminary, do point to informing early childhood professionals about the extent and nature of children's geometrical thinking and 'the richness of children's spontaneous mathematical interactions and the number of geometric aspects that arise in their interaction'. These three papers all consider various facets of mathematization, emphasise the importance of the study of children's interactions and discourse, and use videorecorded recorded data. They set the scene for the remaining papers through their quality, similarities, and diversity.

Papers 4 and 5 in the special issue focus on different aspects of student teachers and how they interact with the mathematics education in their teacher preparation courses. Thiel and Jenßen (Norway) highlight affective aspects of early childhood student teachers' interactions with mathematics. In particular, they investigate the student teachers' mathematical self-efficacy and anxiety and the interaction between these in relation to achievement in their course. The study utilises a strong quantitative approach and is replete with detailed statistical explanations not often seen in the early childhood field. From a methodological aspect, the paper is important because it challenges readers to engage with the quantitative approaches. From an early childhood mathematics education aspect, the study uncovers some unexpected results which question some of the 'conventional wisdom' concerning affect and achievement. An interesting feature of the Thiel and Jenßen paper is the use of student teachers' assessed work as part of the data generation. Similarly, Figueiredo, Gomes, and Rodrigues (Portugal) utilise written assignments from the participant student teachers. In this paper, student teachers responded to a video stimulus by considering both the pedagogical approaches and the mathematical content in the stimulus and suggesting ways of enhancing or continuing the learning shown. The results of this study have some useful consequences for newly developing early childhood teacher education approaches in Portugal, both in terms of the pedagogical content knowledge of prospective early childhood professionals and the specific pedagogical needs of young children. There would appear to be a clear danger that early childhood pedagogies might be subsumed by others seen as appropriate for older children. This could provide yet another example of 'schoolification' in early childhood teacher education.

Notions of representation of mathematical ideas have been canvassed by some of the papers already considered. The next two papers, however, have a specific focus on such representation, although from quite different perspectives. The paper from Northern Ireland authored by Moffett and Eaton reports on the Promoting Early Number Talk (PENT) project and, in particular, the impact perceived by a small group of participating teachers on their own practice and their children's learning about number representation. The paper provides a detailed historical overview of U.K. work from the 1980s on in children's representations of mathematical concepts and thinking and relates this to the impact of a resource book by Casserly, Tiernan, and Moffett (2014) designed to promote early number vocabulary. Findings include that young children's own mathematical representations can provide a 'bridge' between informal and more formal representations and that the valuing and utilisation of children's existing knowledge and skills will best assist children's further learning. Although a relatively 'small' study, this paper does provide a positive stimulus for the approach and further research. Quite a different approach to writing about children's representation is taken by Palmér and van Bommel. They analysed how children in Swedish pre-school classes both represent and systematise their thinking about a combinatorial problem, using Hughes (1986) to classify the representations and Mulligan and Mitchelmore (2009) for the systematisations. The mathematical content of the chosen problem-solving task is quite unusual for early childhood settings, but the study reveals important insights into how children develop abstract thinking. The development of children's systematisations and abstract representations seems to be synchronised, but only if the children's interpretation of the task is taken into account.

Palmér and van Bommel utilise a problem-solving situation in an area of mathematics unfamiliar to most young children (and, incidentally, to many of their educators). In the next paper, Ramírez-Uclés, Castro-Rodríguez, Piñeiro, and Ruiz-Hidalgo (Spain) ask a fundamental question about such situations or, more specifically, the tasks which create a problem-solving situation. Through an analysis of the literature, the authors have determined that real-life tasks suitable for problem-solving with pre-school children should have the following characteristics (wording derived from the paper): The solution is not just a short answer, solvers know who needs the result and why, solving the problem is a multi-stage process, and ideas and procedures from several areas need to be integrated. The findings also show the value of children working in groups to try to solve the problems and the importance of an educator or teacher working with the groups of children in order to stimulate and sustain activity.

As has been intimated earlier, the potential links between play and learning are fertile grounds for investigation in early childhood mathematics education. In the study from Switzerland by Vogt, Hauser, Stebler, Rechsteiner, and Urech, six-year-old children were assigned to one of three 'treatment' groups described as a training program, a play-based approach using card and board games, and a control group. While a detailed analysis of the results is provided in the paper, the major findings are that while the training program benefited children with lower mathematical competency, the play-based approach seemed to benefit all groups of children, regardless of their competency level. As well, the educators preferred the play-based approach, partly because the results reinforced their own beliefs about the appropriateness of the pedagogical approaches.

The final paper for this special issue emanates from the *Let's Count* program (Gervasoni and Perry 2017) as does the earlier paper by MacDonald et al. Perry and Dockett (Australia) use Bronfenbrenner's bioecological framework (Bronfenbrenner and Morris 2006) to analyse responses from early childhood professionals and adult family members about their involvement in the program. There is a particular emphasis on the proximal processes which arise from interactions. Analysis of the data shows that the processes of noticing, exploring and talking about the mathematical activities of pre-school children had major impact on the mathematical attitudes and confidence of the adults involved with these children as well as on the mathematical learning of the children. The paper concludes with recognition from the analysis of the data that 'supporting children's mathematical development involves working collaboratively with those who are in a position to facilitate meaningful, ongoing, regular, reciprocal and increasingly complex interactions with mathematics at their core'.

This special issue of EECERJ is a major achievement of the EECERA SIG Mathematics Birth to Eight Years. We are particularly proud of the diversity of authors, topics, approaches, and countries represented in the collection. Clearly, early childhood mathematics education is an important component of the field and one which engenders much quality research. It is important that such research continues.

References

Bennett, J., and C. P. Taylor. 2006. *Starting Strong II: Early Childhood Education and Care*. Paris: OECD.

Bishop, A. J. 1988. *Mathematical Enculturation: A Cultural Perspective on Mathematics Education*. Dordrecht: Kluwer.

Bronfenbrenner, U., and P. A. Morris. 2006. "The Bioecological Model of Human Development." In *Handbook of Child Psychology, Vol 1: Theoretical Models of Human Development*. 6th ed., edited by W. Damon and R. M. Lerner, 793–828. New York: Wiley.

Broström, S. 2017. "A Dynamic Learning Concept in Early Years' Education: A Possible Way to Prevent Schoolification." *International Journal of Early Years Education* 25 (1): 3–15.

Carmichael, C., A. MacDonald, and L. McFarland-Piazza. 2014. "Predictors of Numeracy Performance in National Testing Programs: Insights from the Longitudinal Study of Australian Children." *British Educational Research Journal* 40 (4): 637–659.

Casserly, A. M., B. Tiernan, and P. Moffett. 2014. *Number Talk: A Resource to Promote Understanding and Use of Early Number Language*. Sligo: The Centre for SEN, Inclusion and Diversity.

Devlin, K. J. 2012. *Introduction to Mathematical Thinking*. Palo Alto, CA: Keith Devlin.

Duncan, G. J., C. J. Dowsett, A. Claessens, K. Magnuson, and A. C. Huston. 2007. "School Readiness and Later Achievement." *Developmental Psychology* 43 (6): 1428–1446.

ECA (Early Childhood Australia), and AAMT (Australian Association of Mathematics Teachers). (2006). *Position Paper on Early Childhood Mathematics*. AAMT Website: http://www.aamt.edu.au/About-AAMT/Position-statements/Early-childhood.

Fröbel, F. 1862. *Die Pädagogik des Kindergartens*. Berlin: Enslin.

Fröbel, F., and I. M. Lilley. 1967. *Friedrich Froebel*. Cambridge: Cambridge University Press.

Geary, D. C., M. K. Hoard, L. Nugent, and D. H. Bailey. 2013. "Adolescents' Functional Numeracy Is Predicted by Their School Entry Number System Knowledge." *PloS One* 8 (1): e54651. PLOS Website: http://journals.plos.org/plosone/article?id = 10.1371/journal.pone.0054651.

Gervasoni, A., and B. Perry. 2017. "Notice, Explore, and Talk About Mathematics: Making a Positive Difference for Preschool Children, Families, and Educators in Australian Communities That Experience Multiple Disadvantages." In *Advances in Child Development and Behavior*, Vol. 53, edited by J. Sarama, D. H. Clements, C. Germeroth, and C. Day-Hess, 169–225. Burlington: Academic Press.

Gruber, K.-H. 2006. "The German 'PISA-Shock': Some Aspects of the Extraordinary Impact of the OECD's PISA Study on the German Education System." In *Cross-national Attraction in Education: Accounts from England and Germany*, edited by H. Ertl, 195–208. Oxford: Symposium Books.

Hardy, G. H. 1940. *A Mathematician's Apology*. Cambridge: Cambridge University Press.

Hemmerling, A. 2007. *Der Kindergarten als Bildungsinstitution: Hintergründe und Perspektiven*. Wiesbaden: VS Verlag für Sozialwissenschaften.

Hughes, M. 1986. *Children and Number: Difficulties in Learning Mathematics*. Oxford: Blackwell.

Katz, L. G. 2010. STEM in the Early Years. *Early Childhood Research and Practice*. ECRP Website: http://ecrp.uiuc.edu/beyond/seed/katz.html.

Moss, J., C. D. Bruce, and J. Bobis. 2016. "Young Children's Access to Powerful Mathematical Ideas: A Review of Current Challenges and New Developments in the Early Years." In *Handbook of International Research in Mathematics Education*. 3rd ed., edited by L. English and D. Kirshner, 153–190. Abingdon: Routledge.

Moss, P. 2014. *Transformative Change and Real Utopias in Early Childhood Education*. London: Routledge.

Mulligan, J., and M. Mitchelmore. 2009. "Awareness of Pattern and Structure in Early Mathematical Development." *Mathematics Education Research Journal* 21 (2): 33–49.

NAEYC (National Association for the Education of Young Children), and NCTM (National Council of Teachers of Mathematics). 2002/2010. *Early Childhood Mathematics: Promoting Good Beginnings*. NAEYC Website: https://www.naeyc.org/sites/default/files/globally-shared/downloads/PDFs/resources/position-statements/psmath.pdf.

Perry, B., and S. Dockett. 2008. "Young Children's Access to Powerful Mathematical Ideas." In *Handbook of International Research in Mathematics Education.* 2nd ed., edited by L. English, 75–108. New York: Routledge.

Samuelsson, I. P., and M. A. Carlsson. 2008. "The Playing Learning Child: Towards a Pedagogy of Early Childhood." *Scandinavian Journal of Educational Research* 52 (6): 623–641.

Sfard, A. 2008. *Thinking as Communicating: Human Development, the Growth of Discourses, and Mathematizing.* Cambridge: Cambridge University Press.

Shen, Y., and C. P. Edwards. 2017. "Mathematical Creativity for the Youngest School Children: Kindergarten to Third Grade Teachers' Interpretations of What It Is and How to Promote It." *The Mathematics Enthusiast* 14 (1): 325–345. University of Montana Website: https://scholarworks.umt.edu/tme/vol14/iss1/19.

von Marenholtz-Bülow, B. 1887. *Handbuch der Fröbelschen Erziehungslehre. Zweiter Teil: Die Praxis der Fröbelschen Erziehungslehre.* Kassel: Georg H. Wigand.

1 Teachers' involvement in children's mathematizing – beyond dichotomization between play and teaching

Camilla Björklund, Maria Magnusson and Hanna Palmér

ABSTRACT

The focus of this article is on mathematics teaching in a play-based and goal-oriented practice, such as preschool, and on how different lines of actions may impact children's learning opportunities. Video recordings of authentic play activities involving children and nine teachers from different preschools were analyzed qualitatively to answer the following research questions: (1) What lines of action do teachers use when they teach mathematics in play? and (2) What implications may different ways of teaching have for children's learning opportunities? The analysis revealed four different categories: confirming direction of interest; providing strategies; situating known concepts; and challenging concept meaning. As these differ regarding both the mathematics content focused on and the kind of knowledge emphasized, they have implications for children's learning opportunities.

Introduction

There is a growing consensus in policy and research that early mathematics is important and bears relevance for children's development in the short and long term. Young children can possess deep and rich mathematical competencies (English and Mulligan 2013; Newton and Alexander 2013), and several studies show that early mathematical competencies have positive effects on later school performance (Duncan et al. 2007; Perry and Dockett 2008; Ginsburg 2009).

Despite the solid view that early mathematics is important, there is no agreement as to how preschool mathematics education should be conducted. Differences in opinion are visible both within and between countries, resulting in a plural view on preschool mathematics (Palmér and Björklund 2016). One of the most prominent differences regards the relation between play and teaching: is teaching to be integrated with or separated from children's play? On the one hand, there are paradigms emphasizing children's right to play, undisturbed by adults, for the sake of play itself (Sundsdal and Øksnes 2015); and on the other hand there is contemporary Nordic research developed within theoretical frameworks that emphasizes a consolidation of the two (Pramling and Pramling Samuelsson 2011; Pramling, Doverborg, and Pramling Samuelsson 2017). The former paradigms often have philosophical underpinnings, as opposed to the latter paradigms' embracing preschool as part of the education system. We find such dichotomies (between play and teaching) unfruitful and contradictory to the fact that many countries around the world include preschool children in the education system and that both teaching and play are central features of this practice.

Sweden is one example whereby early childhood education is available to all children aged one to six years, with a national curriculum that clearly states that the preschool practice is commissioned to ensure that children develop their competencies to their full potential. However, experiences from the Swedish context show that early childhood education is a delicate issue that needs further study (Swedish Schools Inspectorate 2016). As part of the education system, teaching is to be conducted in the Swedish preschool (Education Act 2010:800). At the same time, 'a conscious use of play' (6) is emphasized as important in relation to children's learning (National Agency of Education 2011). What is left for the teachers and researchers within this field to determine is *how* to teach in a play-based and goal-oriented practice. We address this question in this article by studying how teaching (mathematics) in a play-based and goal-oriented practice can be conducted, and how different lines of action may impact children's learning opportunities. The empirical material used is part of a larger research project aiming to investigate the teaching–play relation in Swedish preschool practice. Mathematics as a content for learning is of special interest in this context, since while teachers in general claim that they teach mathematics in preschool (Björklund and Barendregt 2016), an evaluation by the Swedish Schools Inspectorate (2017) shows that mathematics teaching needs to be further developed in a majority of Swedish preschools.

The specific research questions focused on in the article are: (1) What lines of action do teachers use when they teach mathematics in play? and (2) What implications may different ways of teaching have for children's learning opportunities?

Research on mathematics teaching and play

The relation between teaching, mathematics and play can be seen as either 'mathematics made playful', such as games in which counting, sorting and different mathematical operations are prominent, or 'mathematizing elements of play' whereby the primary act is play and a teacher might try to introduce mathematical concepts or operations to the child's play activities (van Oers 1996, 74). The notion of *mathematizing* is often used when the emphasis is on children (including preschool children) trying to understand different phenomena in their surrounding world and mathematics becomes a part of this exploration (Freudenthal 1968). According to this perspective, the mathematics teaching of young children should necessarily be based in the children's own lived experiences (for example play) and involve extending these experiences through mathematical inquiry (Gravemeijer and Terwel 2000). Anghileri (2006, 49) suggests that the teaching process involves the teacher 'initiating reflective shifts such that what is said and done in action subsequently becomes an explicit topic of discussion'. In other words, the actions that children and teacher are involved in become the topic that will be discussed from a mathematical point of view and situated in the activity. This can be related to the act of playing, whereby participants often enter and exit the play context to negotiate the meaning and progress of the ongoing play.

Play activities, and particularly role play, may serve as teaching opportunities if the teacher participates and is able to make use of occurring mathematical phenomena (van Oers 1996). The role of the teacher is then to extend children's encounters with mathematics, in addition to organizing a rich environment that offers opportunities to explore new as well as familiar notions (Wager and Parks 2016). In this approach, it is not enough that mathematical representations and notions are present in the play; the reflection on and enhancement of mathematical thinking in the children depends on the teacher's ability to seize the moment. This comes down to the idea of how children learn; Wager and Parks (2016), as well as van Oers (1996), embrace the idea of children's initiatives and explorations as essential, but emphasize that teachers need to bring in new content and perspectives that will extend the children's experiences, including the children's own play (e.g. Magnusson and Pramling 2017).

This means that teaching mathematics is not merely about promoting counting, adding, naming, or using measures; it is rather about expanding the play and helping the children to understand the surrounding world, thus to mathematize. A key, according to van Oers (1996), can be found in questions from the teacher that encourage the children to discern a problem that emerges in the play activity, helping them mathematize their play content, and furthermore to solve the problem through mathematical operations or representations. There are significant findings on the relation between children's learning outcomes and the frequency and duration of play-based mathematics activities they are engaged in (Cohrssen, Tayler, and Cloney 2015), but less is known about the efficacy of *different ways* of teaching in relation to play.

Theoretical framing

In our study, we base our understanding of teaching on the framework of developmental pedagogy (Pramling and Pramling Samuelsson 2011; Pramling, Doverborg, and Pramling Samuelsson 2017). Similar to the previously described mathematizing (Gravemeijer and Terwel 2000), developmental pedagogy takes its point of departure from children's own lived experiences, whereby teaching implies enabling a child to experience familiar phenomena in new ways or widen his/her experiences, resulting in an extended repertoire of ways to encounter the surrounding world. The teaching triad – consisting of the teacher (who facilitates this extension of experiences), the learner, and the content for learning – is relational. This means that there is a delicate balance between the learner's concept knowledge and the (teacher's) intended way of understanding a concept that relies on their coordinating their perspectives.

Intersubjectivity, in the sense of coordinating perspectives between teacher and learner, is considered necessary in both play and teaching. In play situations, the participants agree on the play rules and boundaries, whereby certain codes are known by the participants and promote the progression of the play. In teaching situations, teacher and learner have to establish a common temporary view, some kind of joint understanding of what they are talking about and that there might be different ways of seeing the object of learning. Thus, a situation in which intersubjectivity is temporarily sufficiently established is a pre-condition for both play and teaching.

In accordance with developmental pedagogy, teaching in early childhood education is understood as supporting children's awareness and their making sense of the surrounding world, which include both getting acquainted with new concepts and experiencing and exploring familiar phenomena in new ways. Support for children's awareness and understanding can be achieved by establishing *sustained shared thinking*. This pedagogical interaction is characterized by 'two or more individuals who "work together" in an intellectual way to solve a problem, clarify a concept, evaluate activities, or extend a narrative' (Siraj-Blatchford 2010, 157). In an interaction in which *sustained shared thinking* is established, both parties (teacher and child/ren) contribute to, develop, and extend their thinking. This kind of interaction has shown to have positive outcomes for children's learning (Siraj-Blatchford 2010), and will work as one guiding principle in our interpretation of implications of teachers' different lines of action.

Methodology and data sources

The study reported on in this article was conducted in collaboration with preschool teachers from different preschools as well as researchers from three universities in Sweden participating in a joint research project. The research relies on authentic documentation from the preschools where children and teachers are engaged in play in different ways and with different content. The legal guardians of the preschool children gave their written consent for their children to participate in video-recorded activities in the regular preschool practice. Information and written forms for participation were distributed by the preschool teachers. Children who were not allowed to be video recorded for the project's purpose have not been observed by the research group or included in the analysis. The names of the participants are fictive in the publications, in order to ensure their integrity.

Data for the current analysis consist of 42 video documentations made by 9 participating preschool teachers. The length of these documentations varies from 3 minutes to approximately 20 minutes. The instructions given to the teachers were to document play activities in which they interact with the preschool children and take part in the children's play to determine what learning objects were possible to develop knowledge about. In the project, these documentations were first analyzed at joint meetings between the researchers and preschool teachers. After these meetings, the observations were transcribed and analyzed in more detail by the researchers, using interaction analysis methods focusing on the dialogue between teacher and child/ren and their joint construction of the activity (Bryman 2008).

The guiding question in this analysis was how teaching is framed in the play activities and what became possible for the participating children to learn. In this article, we direct specific attention to play in which the children mathematize and the teachers act to support the children's awareness and understanding, in accordance with developmental pedagogy principles. The results presented in the next section are based on 42 video documentations, of which 30 contained mathematical content. We analyzed the activities, focusing on the ways the teachers established sufficient intersubjectivity in order to find the different ways in which teachers participate and teach in play. We found four different lines of action used by the teachers. The unit of analysis is the dialogical interaction about mathematical content, rather than the teachers themselves. This means that several actions were found within the same video recording, and the preschool tea-chers varied their lines of action on different occasions. Thus, the categorization should not be understood quantitatively; instead; the purpose has been to explore *different ways* of teaching mathematics in relation to play. Our analysis shows that no single line of action dominates over another in our sample; they occur fairly evenly in the play activi-ties. In the next section, we will present the four lines of actions, illustrated with examples from two play activities with two different preschool teachers. Our choice to use these two activities derives from the finding that different lines of action are used within the same play by the same preschool teacher. Nevertheless, together these examples illustrate the four re-occurring lines of action found in the video documentations.

Results

Four different lines of action emerge in our analysis: *confirming direction of interest*; *pro-viding strategies*; *situating known concepts*; and *challenging concept meaning*. In the follow-ing presentation, these are illustrated with examples from the analyzed observations. Our analysis reveals how the teacher and children initiate and respond to the mathematical content present in their dialogue, and which opportunities for mathematics learning emerge.

Confirming direction of interest

The first line of action – confirming direction of interest – implies that the teacher shows interest in what the children are engaged in by asking questions or repeating a child's utterance. Confirming or repeating what a child says or does is a common way for teachers to interact with children. In this way, they express that they are aware of the child's directed interest, and usually that they support the ongoing activity.

Below is an example of *confirming direction of interest*. In the example, we find three children (aged four and five) and one teacher sitting around a table. On the table is a large box of building material. The children are making Beyblades (spin toys connected to a popular game and TV series in which the players compete with their spinning Bey-blades) with the material. At the start, one child is spinning one of his Beyblades. The other children and the teacher observe how the color of the Beyblade changes as it spins.

Excerpt 1: Spinning Beyblades

1.	Teacher:	It [referring to the color] changes a little bit, anyway. How did that happen? That it turned blue?
2.	Mark:	I don' t know. But. Because I have magic fingers.
3.	Teacher:	You have magic fingers. Yes, that' s an idea.
4.	Tom:	Now I' m going to make a panda.
5.	Teacher:	You' re going to make a panda.

The question of why the Beyblade changes color as it spins can be regarded as possible content for learning within the frames of the ongoing play. An open question like this can become the basis for *reasoning*, as in using arguments to motivate one's answer. Thus, with her question 'How did that happen?' the teacher introduces two possible learning objects, the phenomenon of the changed color and the competence of reasoning. Reasoning is emphasized as an important mathematical competence (Kilpatrick, Swafford, and Findell 2001), even in early mathematics (National Agency for Education 2011) and is commonly found in similar kind of interactions in early childhood education. The child answers that the color changes because he has 'magic fingers' and thus produces a plausible explanation for the phenomenon. According to Lithner (2008), reasoning does not have to be grounded in formal logic and the arguments used can even be considered incorrect from an outsider's perspective, as long as they make sense and are reasonable to the individual in the situation.

Questions like the one posed by the teacher in Excerpt 1 make it possible to explore reasoning, which can then gradually be developed into mathematical reasoning. The open character of the posed question promotes *creative reasoning* whereby the child has to develop new (to him) arguments as he does not know the formal answer to the question (Lithner 2008). However, for this to occur the child's argument (magic fingers) would need to be challenged by the teacher, encouraging the child to develop his argument. However, the continuation of the dialogue does not focus on the change of color; instead, the teacher starts interacting with another child. In the continuation of the activity the children build and spin Beyblades, and the teacher confirms what the children are saying and asks new questions.

A common characteristic of this way of confirming children's directed interest is the teacher repeating the children's initiatives and thereby confirming the notions they use, whether it is a search for the smallest pig that has suddenly gone missing in a fairytale told with props, or the sorting of colors to make houses for play figures. The confirmation of notions used as proper for the current situation is one important aspect of mastering known mathematical notions. Nevertheless, in this strategy, these notions are not challenged or extended to support further concept development.

Providing strategies

The next line of action found is the teacher providing children with strategies. Many teachers seem to focus their pedagogical efforts on supporting children in mastering counting skills, which in our study seems to become a challenging skill to master for many of the participating children. When a teacher provides a strategy it is done by supporting the children in pointing while counting, directing attention to what has been counted or reciting the counting rhyme out loud along with the children. These are considered necessary strategies for solving a counting task, and direct the children's actions towards accessible paths.

The example below of *providing strategies* is from the same play activity as the previous example. Later in the same building activity one of the children is counting her Beyblades. The girl counts out loud, touching one Beyblade at a time. However, as she counts she does not match her counting to each individual item, mixing which Beyblades she has and has not counted.

Excerpt 2: Counting slowly

10.	Teacher:	If you count slowly Sarah. Then, then. Do it from the beginning and then count, yes.
11.	Sarah:	One, two, three, four, five, six, seven. (She counts out loud as she moves one Beyblade at a time to the side on the table)
12.	Teacher:	How nice. Yes, it' s always good to count slowly.

The teacher has a clear goal for which skill she wants the girl to develop: counting. The coordination of number words and objects, whereby one and only one number word is assigned to each object, is one necessary how-to-count principle (Gelman and Gallistel 1978). At the same time, not being able to coordinate verbal counting words with actions such as pointing at objects is common among young children (Clements and Sarama 2009). Thus, counting slowly and moving one item at a time are reasonable strategies for supporting this child's counting act. In this situation, the teacher makes visible and further develops a mathematical procedure occurring in the play. However, while this is being done, the activity moves out of the play frame towards an instructive frame whereby the teacher provides strategies to facilitate her counting. It is not obvious how counting slowly – or counting in the first place – is part of the building of or playing with the Beyblades. Other counting acts found in the analyzed observations also include the counting of objects as part of the play, for instance when the teacher and children go to Antarctica with their many animals and all of them have to be on board. These counting tasks are most often introduced by the teachers, but provide the children with opportunities to mathematize the content of their play.

Situating known concepts

A third line of teachers' action in play appears when the children, or the teacher, use familiar concepts to act in the play, but a joint exploration of the meaning of the concept occurs when someone (the teacher or a child) takes initiatives to situate the concept within the play. The concept then adds meaning *to the play*, and the concept meaning is situated in the actual play.

In the example (Excerpt 3) of *situating known concepts*, we find two children and one teacher engaged in role play in a separate room at the preschool. The teacher plays the character of a younger sister. The two children, Lisa and Alice, act as older sisters. Together they are planning for the younger sister's birthday party, and of particular importance is the concept of space, which is situated in the play frame.

Excerpt 3: Planning a birthday party

1.	Teacher:	I would like to invite Kalle and Lisa and Olle. (holds up one hand, extending one finger for each name she says) How many can I invite? Ten, twenty?
2.	Lisa:	As many as you'd like.
3.	Teacher:	Twenty, Thirty, ah! Where should we be then? In my room?
4.	Lisa:	At home.
5.	Teacher:	At home? How big is it then?
6.	Lisa:	Big!
7.	Teacher:	Ha!
8.	Alice:	This big! (Alice stretches her arms to each side)

Both counting and space are present in this example, but as the dialogue turns out, 'space' can be understood as the primary content. While acting in her role as little sister, the teacher makes space visible as a mathematical phenomenon. By the teacher asking 'how big is it then' (referring to the size of the imagined house), space is discussed in relation to how many to invite, which can be regarded as an example of functional measurement (Clements and Sarama 2009). However, since the notion of space is explored in relation to an imagined home, the space is not physically present and cannot be measured in a normative way. This is an example of a teacher's questions encouraging children to discern a problem emerging from the play activity. Her question invites the children to mathematize their play content, and furthermore to solve the problem through mathematical operations and representations (van Oers 1996). They share a common interest in figuring out how many guests to invite in relation to the available space, which is visualized through different expressions such as finger counting and stretched arms. The reasoning by Lisa and Alice is mathematical, referring to space in both wording (big) and gestures (arms).

As seen in the example above, the reasoning about known concepts is not necessarily creative but rather imitative (Lithner 2008), as the children use notions and procedures they are already familiar with and no new mathematics are explored in the situation. From this line of action, what the children are offered to learn is an extension of the known concept, which is of importance in preschool where many mathematical phenomena are necessarily situated in concrete situations (see Björklund 2018) and in meaningful use (such as play).

Challenging concept meaning

The fourth line of action we found is when teachers use play activities to become engaged in exploring concept meaning together with the children. What stands out in this way of teaching mathematics in play is that the teacher does not provide an answer, or strategies, but rather contrast meanings to what the children express. In this way, a concept emerging in the play is highlighted and the children are challenged and inspired to explain their view and elaborate their expressions of meaning, in order to establish a shared (and more advanced) understanding.

The example below of *challenging concept meaning* is from the same role play as the previous example (Excerpt 3). In this, the teacher and the two children are talking about their ages in the role play. Talking about the ages can be considered meta-communication about the play, but the teacher continues to act as the little sister while they are talking. (Important in relation to the dialogue below is that, in Sweden, the year before compulsory school begins is commonly called 'the zero'. This refers to the grade before 'first grade'/'the one', when children start compulsory school at the age of seven.)

Excerpt 4: The meaning of zero

44.	Alice:	Aa … I'm seven years and I'm in zero … in first grade.
45.	Teacher:	So you're in first grade?
46.	Alice:	You'll start in the zero.
47.	Teacher:	Zero! But that's nothing! She says that I'll start in the zero.
48.	Lisa:	They're next to us.
49.	Alice:	The zero **in school**!
50.	Teacher:	Is there a zero in school?
51.	Alice:	Yes, there's a zero.
52.	Teacher:	Aha.
53.	Alice:	Yes, that's zero and I'm in the one.
54.	Teacher:	Yes, and you're in the one.

Throughout the example, the teacher is acting in her role as little sister and the children in their roles as big sisters. In the dialogue, 'zero' becomes an object of exploration with the teacher highlighting the possibility to interpret the expression in different ways. Numbers imply different things in different situations, and by saying 'Zero! But that's nothing!' the teacher makes visible two dimensions of how the numeral zero can be used, as the identification of a grade in school and as a quantity. Preschool children who lack formal mathematics education often have a limited understanding of zero, since at this age counting is often connected to quantity and it seldom makes sense to count zero objects (Clements and Sarama 2009). In this example, the teacher and children are working together in an intellectual way to solve the problem of what zero implies, and the children contribute to the explanation by saying 'The zero **in school**' and 'Yes, that's zero and I'm in the one'. The teacher's role is of certain importance, since the two dimensions of 'zero' would not have been challenged if the teacher had not brought in the contrasting meaning. This way of teaching concept meaning appears quite often in the role play in our study, with the teacher introducing new concepts or a different meaning of a concept in a natural way.

Conclusions

The starting point for this article was the ongoing debate regarding the relation between play and teaching, and the question of how to teach (mathematics) in a play-based and goal-oriented practice (Pramling and Pramling Samuelsson 2011; Sundsdal and Øksnes 2015; Pramling, Doverborg, and Pramling Samuelsson 2017). In the results, through the use of examples, we have illustrated four different lines of action in teaching mathematics in play: *confirming direction of interest*; *providing strategies*; *situating known concepts*; and *challenging concept meaning*. In the examples, the relation between mathematics and play is what van Oers (1996) refers to as mathematizing elements of play; implying that mathematical concepts or operations that can contribute to the play are explored and/or introduced. Our interest is directed towards the teachers' involvement in the children's mathematizing within the frames of play. Different ways to respond to children's mathematizing initiatives constitute different learning opportunities, in which the teacher's responsiveness to the children's acts and understanding is a key feature. This includes both retaining intersubjectivity and handling the delicate balance between remaining within the play frame and extending the children's experiences. In this section, we will conclude our results and note their similarities and differences.

Our analysis shows that the outcome of the teaching act turns out differently depending on the teachers' responsiveness (how they interact/respond) to the children's mathematizing. The four lines of action have in common that opportunities occur in which it is possible to challenge the conceptual meaning of mathematical notions and in different ways extend the children's experiences. While some possible mathematical content for learning is thereby a necessary starting point, this possible mathematical content for learning can be dealt with in different ways. Different lines of action will open up for different opportunities for the children to deal with mathematical questions. To *confirm direction of interest* means that the teacher is responsive to the children's ideas; however, this confirmative approach seldom seems to benefit an extension of knowledge or inquiry of the phenomenon that was initiated. *Providing strategies*, on the other hand, is a more goal-oriented act of directing the child towards skills or tools that will help him/her master a challenge. This strategy can be associated with the notion of guided participation (Rogoff 1990), which includes interpersonal actions whereby the teacher and learner go side-by-side in a culturally organized activity. The key to the adult's role here is being responsive to the child's perspective, rather than directing it. Thus, mathematics teaching in line with this can be understood as the teacher providing the child with new strategies to facilitate the activity the child is involved in. *Situating known concepts* implies mathematical content being applied to a current play context. The content may come to be of importance for the progression of the play – and indeed, making connections between and generalizing the use of mathematical concepts is part of what mathematics is about (Anghileri 2006). Finally, *challenging concept meaning* includes sustained shared thinking, whereby the teacher and child/ren work together in an intellectual way to solve a problem or clarify a concept (Siraj-Blatchford 2010). The key in this line of action is the teacher introducing some new meaning or perspective that extends the child/ren's experiences and brings forth a new way of understanding or adds value to the play.

Preschool mathematics is not (only) about pre-prepared activities but also entails teachers being involved in children's mathematizing activities, for example in play (Siraj-Blatchford and Sylva 2004). Based on our study, we cannot see a dichotomy between play and teaching as outlined by, for example, Sundsdal and Øksnes (2015). Their interpretation, based on a narrow interpretation of teaching, results in a false dichotomy between play and teaching, which is not fruitful for our understanding of either play or teaching in preschool. Our results, presented in this article, show that the activity of play continues even when the teacher clearly teaches mathematical content. The teachers can offer extensions without destroying, disrupting or controlling the play. Instead, the teachers more or less extend the mathematical content within the play in different ways. The three lines of action *providing strategies*, *situating known concepts* and *challenging concept meaning* are in line with developmental pedagogy (Pramling Samuelsson and Asplund Carlsson 2008), in that the teacher supports the children's awareness and exploration of meaning, either by exploring familiar phenomena in new ways or by helping them get acquainted with a new or contrasting meaning of a phenomenon. In the case of *challenging concept meaning*, a key feature of teaching in a play-oriented practice is highlighted: balancing between the challenging contrasts while still maintaining the intersubjectivity and play frame that have been established. Regardless of strategy, the teachers' responsiveness to the children's ideas is likely the most essential feature in the play and teaching activities (cf. Wager and Parks 2016); but in order for learning to occur (in the sense of extending one's knowledge and understanding) there are also other features that seem necessary, such as a joint problem to solve or stated differences in the perception of a common phenomenon. In line with van Oers (1996), we also find that teachers' questions form a key feature of preschool mathematics teaching, in that they frame the opportunities children are given to discern such problems that emerge in the play activity, helping them mathematize their play content.

Based on the results presented, we do not claim that mathematics teaching in preschool should occur only in play, but we do claim that the mathematizing of elements of play is both a possible and desirable part of mathematics teaching in preschool. These results are important as a contribution to the current debate about the policy and practice of teaching in early childhood education. A false dichotomy between play and teaching carries the risk that the focus of discussion will be on the wrong question, asking *if* instead of *how*.

Disclosure statement

No potential conflict of interest was reported by the authors.

Funding

The study is funded by the Swedish Institute for Educational Research, project nr. 2016/112.

ORCID

Maria Magnusson ⓘ http://orcid.org/0000-0002-5806-4475

References

Anghileri, J. 2006. "Scaffolding Practices that Enhance Mathematics Learning." *Journal of Mathematics Teacher Education* 9 (1): 33–52. doi:10.1007/s10857-006-9005-9.

Björklund, C. 2018. "Learning About the Notion 'Half': Critical Aspects and Pedagogical Strategies in Preschool." *Scandinavian Journal of Educational Research* 62 (2): 245–263. doi:10.1080/00313831.2016.1212264.

Björklund, C., and W. Barendregt. 2016. "Teachers' Mathematical Awareness in Swedish Early Childhood Education." *Scandinavian Journal of Educational Research* 60 (3): 359–377. doi:10.1080/00313831.2015.1066426.

Bryman, A. 2008. *Social Research Methods*. 3rd ed. Oxford: Oxford University Press.

Clements, D. H., and J. Sarama. 2009. *Learning and Teaching Early Math. The Learning Trajectory Approach*. New York: Routledge.

Cohrssen, C., C. Tayler, and D. Cloney. 2015. "Playing with Maths: Implications for Early Childhood Mathematics Teaching from an Implementation Study in Melbourne, Australia." *Education* 43 (6): 641–652. doi:10.1080/03004279.2013.848916.

Duncan, G. J., C. J. Dowsett, A. Claessens, K. Magnuson, A. C. Huston, P. Klebanov, L. S. Pagani, et al. 2007. "School Readiness and Later Achievement." *Developmental Psychology* 43 (6): 1428–1446.

English, L. D., and J. T. Mulligan. 2013. "Perspectives on Reconceptualizing Early Mathematics Learning: Introduction." In *Perspectives on Reconceptualizing Early Mathematics Learning*, edited by L. D. English and J. T. Mulligan, 1–4. Dordrecht: Springer.

Freudenthal, H. 1968. "Why to Teach Mathematics so as to Be Useful" *Educational Studies in Mathematics* 1 (1): 3–8.

Gelman, R., and C. R. Gallistel. 1978. *The Child's Understanding of Number*. London: Harvard UP.

Ginsburg, H. P. 2009. "Early Mathematics Education and How to Do It." In *Handbook of Child Development and Early Education: Research to Practice*, edited by O. A. Barbarin and B. H. Wasik, 403–428. New York: Guilford Press.

Gravemeijer, K., and J. Terwel. 2000. "Hans Freudenthal: A Mathematician on Didactics and Curriculum Theory." *Journal of Curriculum Studies* 32 (6): 777–796.

Kilpatrick, J., J. Swafford, and B. Findell. 2001. *Adding It Up: Helping Children Learn Mathematics*. Washington, DC: National Academies Press.

Lithner, J. 2008. "A Research Framework for Creative and Imitative Reasoning." *Educational Studies in Mathematics* 67 (3): 255–276.

Magnusson, M., and N. Pramling. 2017. "In 'Numberland': Play-Based Pedagogy in Response to Imaginative Numeracy." *International Journal of Early Years Education*. doi:10.1080/09669760.2017.1368369.

National Agency for Education. 2011. *Curriculum for the Preschool Lpfö98*. Revised 2010. Stockholm: National Agency for Education.

Newton, K. J., and P. A. Alexander. 2013. "Early Mathematics Learning in Perspective: Eras and Forces of Change." In *Perspectives on Reconceptualizing Early Mathematics Learning*, edited by L. D. English and J. T. Mulligan, 5–28. Dordrecht: Springer.

Palmér, H., and C. Björklund. 2016. "Different Perspectives on Possible – Desirable – Plausible Mathematics Learning in Preschool." *Nordic Studies in Mathematics Education* 21 (4): 177–191.

Perry, B., and S. Dockett. 2008. "Young Children's Access to Powerful Mathematical Ideas." In *Handbook of International Research in Mathematics Education*, edited by L. D. English, 75–108. New York: Routledge.

Pramling, N., E. Doverborg, and I. Pramling Samuelsson. 2017. "Re-metaphorizing Teaching and Learning in Early Childhood Education Beyond the Instruction – Social Fostering Divide. International Perspectives on Early Childhood Education and Development." In *Nordic Social Pedagogical Approach to Early Years*, edited by C. Ringsmose and G. Kragh-Müller, 205–218. Switzerland: Springer.

Pramling, N., and I. Pramling Samuelsson, eds. 2011. *Educational Encounters: Nordic Studies in Early Childhood Didactics.* Dordrecht: Springer.

Pramling Samuelsson, I., and M. Asplund Carlsson. 2008. "The Playing Learning Child: Towards a Pedagogy of Early Childhood." *Scandinavian Journal of Educational Research* 52 (6): 623–641.

Rogoff, B. 1990. *Apprenticeship in Thinking: Cognitive Development in Social Context.* Oxford: Oxford University Press.

Siraj-Blatchford, I. 2010. "A Focus on Pedagogy: Case Studies of Effective Practice." In *Early Childhood Matters. Evidence from the Effective Pre-school and Primary Education Project*, edited by K. Sylva, E. Melhuish, P. Sammons, I. Siraj-Blatchford, and B. Taggart, 149–165. London: Routledge.

Siraj-Blatchford, I., and K. Sylva. 2004. "Researching Pedagogy in English Pre-schools." *British Educational Research Journal* 30 (5): 713–730.

Sundsdal, E., and M. Øksnes. 2015. "Til forsvar for barns spontane lek." *Nordisk tidskrift for pedagogikk og kritikk* 1: 1–11.

Swedish Schools Inspectorate. 2016. *Förskolans pedagogiska uppdrag* [Preshools' Pedagogical Mission]. Stockholm: Swedish Schools Inspectorate.

Swedish Schools Inspectorate. 2017. *Förskolans kvalitet och måluppfyllelse* [Preschool Quality and Aims Fulfilled]. Stockholm: Swedish Schools Inspectorate.

van Oers, B. 1996. "Are You Sure? Stimulating Mathematical Thinking During Young Children's Play." *European Early Childhood Education Research Journal* 4 (1): 71–87. doi:10.1080/13502939685207851.

Wager, A., and A. N. Parks. 2016. "Assessing Early Number Learning in Play." *ZDM Mathematics Education* 48 (7): 991–1002. doi:10.1007/s11858-016-0806-8.

2 Young children's mathematical learning opportunities in family shopping experiences

Amy MacDonald, Angela Fenton and Christina Davidson

ABSTRACT

This article reports on a qualitative pilot study which documented the ways in which young children and their families *notice*, *explore*, and *talk about* mathematical concepts and processes as they participate in family shopping experiences. Six families, with children ranging from 12 months to 10 years, were video- and audio-recorded whilst shopping at 1 of 2 large retailers. The data reveal that young children and their families notice, explore, and talk about a great deal of implicit and explicit mathematics whilst shopping together. All of the children displayed instances of mathematical noticing, with the children 'marking' what they had noticed in both verbal and non-verbal forms. Furthermore, all six families explored and talked about what was noticed whilst shopping together. This study contributes new knowledge about the ways in which children and families interact with mathematics in community contexts.

Introduction

It is well established that young children learn a great deal of mathematics in their early childhood years. Everyday contexts, outside of formal learning environments such as schools and early childhood centres, provide opportunities for meaningful explorations of mathematics (Guberman 2004). Family shopping experiences present one such context. This article reports on a pilot project which aimed to use a range of innovative digital recording techniques to examine young children's opportunities for learning about mathematics whilst shopping with their family. The extant literature surrounding mathematical learning in the early years is primarily focused upon learning in either home or early childhood centre/school contexts. This project seeks to expand the focus of mathematical learning in the early years to place greater importance upon the learning and interactions which occur in community contexts outside of the home, early childhood centre, or school. This article investigates the ways in which everyday activities such as shopping provide powerful opportunities for mathematical learning and interactions in community contexts, and reinforces that significant mathematical learning occurs in the early childhood years. Specifically, this article explores the following research questions:

(1) What mathematics can be noticed in family shopping experiences? (2) In what ways is it noticed, and by whom?, and (3) In what ways do children and families explore and talk about mathematics while shopping together?

Background

Children begin developing mathematical skills from a very young age. International research has shown that babies and toddlers demonstrate competence in regard to a range of mathematical concepts and processes, including number and counting, geometry, dimensions and proportions, location, and problem-solving (Björklund 2008; Reikerås, Løge, and Knivsberg 2012). Both Australian and international research has established that young children engage with a range of mathematical concepts and processes prior to starting school (e.g. Gervasoni and Perry 2015; Sarama and Clements 2015). The seminal Australian study, the *Early Numeracy Research Project* (see, for example, Clarke, Clarke, and Cheeseman 2006) investigated the mathematical knowledge of over 1400 children in their first year of primary school. An important finding from the study was that much of the content which formed the mathematics curriculum for the first year of school was already understood clearly by many children on arrival at primary school (Clarke, Clarke, and Cheeseman 2006), a finding echoed in several other studies, both in Australia (for example, Gervasoni and Perry 2015; MacDonald 2010) and internationally (for example, Aubrey 1993; Wright 1994). Research has emphasised the importance of this early mathematical learning, with links being drawn between early mathematics and later achievement (MacDonald and Carmichael 2016; Watts et al. 2014). This research notes, in particular, the predictive power of mathematical knowledge at school entry for later mathematical achievement (Duncan et al. 2007). It has been found that children who enter primary school with high levels of mathematical knowledge maintain these high levels of mathematical skill throughout, at least, their primary school education (Baroody 2000; Klibanoff 2006). Furthermore, De Lange (2008) has suggested that in the years prior to commencing formal education, young children have a curiosity about scientific phenomena, including mathematics. In an Australian study of teacher-reported data for 6500 children, MacDonald and Carmichael (2015) found that 98% of the children showed interest in numbers at 4–5 years. If children engage in meaningful and enjoyable mathematics education in the early childhood years, they are much more likely to appreciate and continue to engage in later mathematics education (Linder, Powers-Costello, and Stegelin 2011).

Everyday community contexts provide powerful opportunities for learning. For example, studies into young children's visits to community spaces such as museums, art galleries, and zoos have demonstrated the potential for learning in such contexts (for example, Fasoli 2003; Tunnicliffe 1995). In mathematics education, the work of Nunes, Schliemann, and Carraher (1993) demonstrated the power of everyday uses of mathematics (what they termed 'street mathematics') for children's mathematical thinking and learning, giving the example of children's participation in street markets. Worthington and Carruthers (2003) argued that family activities such as shopping, cooking, and household tasks provide many opportunities for exploring mathematics, including a variety of calculations, all aspects of measurements, probability, and money. More recently, MacDonald's (2012) research with young children revealed that children themselves identify community contexts such as shopping centres and supermarkets as places in which they notice and learn about mathematics. Community contexts also provide opportunities for exploration of environmental print which may be rich in mathematical concepts. Environmental print encompasses 'labels, signs and other kinds of print' (Horner 2005) found in everyday contexts. Interactions with parents and teachers are considered integral to children's learning about environmental print (Neumann et al. 2011). However, existing studies foreground interactions in the home and at school. As such, family interactions with environmental print *outside of home and school* are the focus in this study.

An emphasis on everyday activities such as shopping represents a strengths-based view of the learning resources available to children and families. The *strengths approach* (McCashen 2005) recognises that all children and families have resources upon which to draw to support their learning and development, and shopping is a context that provides such a resource. Research into the use of a strengths approach has yielded positive results for recognising the resources of children, families, and communities (McCashen 2005; Saleebey 2009). Anderson (1998) has advocated for a focus on acknowledging and highlighting the role that families play in their children's early mathematics learning and has suggested that families and educators can assist one another to recognise children's mathematical potential. This pilot project draws on current research (Fenton 2013) confirming that strengths approaches can contribute to improved educational outcomes, and makes visible the mathematical learning resources available to children and families in everyday contexts such as shopping.

Theoretical framework

One way of recognising the mathematical learning opportunities in everyday contexts such as shopping is by attending to what children and families notice, explore, and talk about (Perry and Gervasoni 2012) as they shop. Developed for the *Let's Count* early numeracy programme, the mantra 'notice, explore, and talk about mathematics' (Perry and Gervasoni 2012) provides a useful framework for attending to the activities and interactions of children and families.

The *Let's Count* mantra builds on the work of Mason (2002) in relation to 'intentional noticing', which Mason describes as processes of 'noticing, marking/remarking, and recording'. As Mason explains:

> To notice is to make a distinction, to create foreground and background, to distinguish some 'thing' from its surroundings. This may not be conscious … Thus *to notice* can be taken to mean the same as *to perceive*, even *to sense* in the most general 'sense' of that word. (Emphasis in the original, 33)

Mason suggests that it is useful to distinguish between *noticing* and *marking*, 'in which not only do you notice but you are able to initiate mention of what you have noticed' (33). Furthermore:

> *Marking* signals that there was something salient about the incident, and re-marking about it to someone else or even yourself makes the incident more likely to be available for yet further access, reflection and re-construction in the future. Thus *marking* is a heightened form of noticing. (Emphasis in the original, 33)

The third element of Mason's framework describes the desire to note or *record* in some way what has been noticed. This may be through the production of text (for example, a list), but it is also possible to note something inwardly, making a *mental note* and initiating a state in which you might choose to re-mark or note outwardly at some future moment (34).

Taken together, these frameworks of 'noticing' provided conceptual structures to guide analysis of the mathematical learning opportunities recorded within the data.

Mathematical framework

Bishop (1988) identified six 'universal mathematical activities' – counting, measuring, locating, designing, playing, and explaining – which, he argued, can be identified in the everyday practices of people across varying cultures and contexts. This framework of 'mathematical activities' was used to conceptualise the everyday explorations of mathematics which were the focus of this study. Macmillan (2009, 245) summarises Bishop's six mathematical activities as follows:

- *Counting* – expressions of numerical quantifiers and qualifiers, including symbolic attributions indicating understanding, awareness, and appreciation of quantity.
- *Measuring* – expressions of non-numerical quantities or qualities.
- *Locating* – expressions of position, shape, boundedness, continuousness, direction, physical, or temporal space.
- *Designing* – expressions of a symbolic plan, structure or shape, or surface of a shape.
- *Explaining* – expressions of factual or logical aspects of ideas, questions, experiences, events, or relationships between phenomena.
- *Playing* – expressions of imaginative or imitative recreation of social, concrete, or abstract models of reality.

Bishop's framework is suitable for exploring families' mathematical interactions through its focus on the *activities* of mathematics within a range of cultural contexts.

Research design

The aim of the pilot study was to trial and evaluates methods of documenting the mathematics that families with young children notice, explore, and talk about (Perry and Gervasoni 2012) as they participate in family shopping experiences. The study aimed to document the ways in which families interacted about mathematics in community contexts and the learning opportunities for children this interaction affords.

Participants

A convenience sample of families with young children was invited to participate in this pilot study. Families were invited on the basis that they had one or more children aged within the range of birth to eight years. The invitation to participate did not specify a focus on mathematics; rather, families were informed that the researchers were interested in documenting the things families with young children notice, explore, and talk about, generally, as they participate in family shopping experiences. Retailers were invited on the basis that they provided a large, family-friendly store environment suitable for shopping with a trolley. Six families and two retailers from two regional Australian towns agreed to participate in the study (three families per store, per town).

Family participants

Details of the six families who participated in this project are presented in Table 1. Pseudonyms have been applied to all participants.

Store participants

Two large retailers, each in one of two large, regional Australian towns, participated in this project. Both stores were locally owned businesses that had been in their respective communities for several decades, and were regular shopping venues for the families in this project. Store 1 was a combined fruit and vegetable retailer and butcher; whilst Store 2 was a combined butcher and delicatessen. Both stores provided large-format, family-friendly shopping environments suitable for navigating with a shopping trolley. Furthermore, both stores provided shopping environments rich with diverse products, displays, infographics, and environmental print. It was hypothesised that these store environments would present a variety of everyday, naturalistic opportunities for children and families to notice, explore, and talk about mathematics whilst shopping.

Ethical considerations and consent processes

This project was reviewed and approved by the Faculty's Human Research Ethics Committee (approval no. 100/2016/06). Extensive consideration was given to the ethical issues of recording children and families in public places. Families were provided with a detailed information sheet about the project, and also participated in several conversations with the researchers prior to making a decision about their involvement.

Children's participation was at the discretion of their parents. No child was able to participate without their parent present. Once informed consent was given by the parents, the project was discussed with the children and they too were given the opportunity to assent to their participation. The researchers also discussed with parent participants that if at any point their children signalled that they did not wish to participate in the data gathering

Table 1. Family participants.

Site no.	Family no.	Parent participant	Child participant/s
1	1	Mother: Karen	Daughter: Evie, 23 months
	2	Father: Corey	Son: Jasper, 22 months
	3	Father: Brett	Son: Damian, 10 years
			Son: Nelson, 8 years
			Son: Cash, 6 years
			Daughter: Rebecca, 3 years
2	4	Mother: Sandie	Son: Zack, 8 years
			Son: Callum, 7 years
			Daughter: Ava, 3 years
	5	Mother: Jillian	Daughter: Tilly, 3 years
			Son: Leo, 12 months
	6	Mother: Lucy	Daughter: Isobel, 6 years

activity (for example, exhibited through their behaviour or a verbal indication), then their indication of dissent should be respected and data gathering should cease.

Data gathering methods

Families were invited to undertake a normal shop at the store in their respective town. The families each undertook their shop at different times. During their shopping experience, the activities and interactions of the family members were captured via a range of innovative digital recording methods. The data gathering techniques were designed to be minimally intrusive so as to preserve the naturalistic nature of the activity and interactions. All families were invited to complete their shopping using a shopping trolley mounted with a custom-built Go-Pro© camera rig (nicknamed 'trolley-cam' for this project). The trolley-cam was specially designed to capture the interactions of the family with the store environment, within close range (approximately arms' length); rather than filming the wider store environment or other shoppers. The trolley-cam was mounted at the front of the trolley and directed back towards the parent pushing the trolley and the child in the trolley seat, so as to capture the interactions of the parent and child. The parent was also fitted with a lapel-microphone voice recorder to ensure that high-quality audio recordings of parent–child interactions were captured. In family groups with older children, the children were invited to wear a pair of glasses with an in-built camera to capture point-of-view recordings. The recorder glasses closely resembled an ordinary pair of eye-glasses and provided an inconspicuous means of recording video data. Two children consented to recording their participation in this way.

The researchers were present to assist the families with the recording equipment prior to, and at the conclusion of, the shopping. The families completed their shopping unaccompanied by the researchers; however, the researchers remained on stand-by at a visible and accessible location within the store throughout the data gathering process. Families were not given any instructions as to how to undertake their shop or what to purchase, and the recording ceased at the point of sale. The length of the recorded shopping trips ranged from 8 to 22 minutes.

Data analysis

A number of analytic techniques were used to interrogate the activities and interactions captured during the family shopping experiences. Bishop's Mathematical Activities (1988) was used as a coding frame for identifying and labelling the mathematical concepts and processes noticed in the data. Data were coded separately by the researchers before comparing codes to ensure consistency. Conversational analyses (Davidson 2010), and gestural analyses (McNeill 1992) attended to the ways in which the families interacted about the mathematics that was noticed. Finally, theoretical thematic analysis of the multimodal data was used to identify key constructs and patterns within the data (Joffe 2011). Theory-building was guided by Mason's (2002) 'intentional noticing' framework along with Perry and Gervasoni's (2012) 'notice, explore, talk about' adaptation for young children's learning in family contexts.

Results

In this section, a selected vignette from each of the six families is presented. Each vignette is analysed in relation to Bishop's Mathematical Activities (1988) to demonstrate the ways in which the families noticed, explored, and talked about mathematics whilst shopping.

Family 1

Family 1 mother, Karen, is shopping with her daughter, Evie (23 months). Karen is pushing a trolley in which Evie is seated. Although the seat is facing Karen, Evie positions herself to face forward, looking in the direction that the trolley is moving, rather than back towards her mother. In this vignette (Table 2), Karen and Evie have just entered the store and are beginning their shopping.

Family 2

Family 2 father, Corey, is shopping with his son, Jasper (22 months). Corey is pushing a trolley in which Jasper is seated, facing Corey. In this vignette (Table 3), Corey and Jasper are moving through the fruit section of the store.

Family 3

Family 3 father, Brett, is shopping with his four children: son, Damian (10 years); son, Nelson (8 years); son, Cash (6 years); and daughter, Rebecca (3 years). Brett is pushing a trolley in which Rebecca is seated, and Nelson is walking beside them. The two older boys, Damian and Nelson, are pushing a second trolley. Brett sends Damian and

Table 2. Vignette from Family 1.

Observations	Mathematical activities
Karen (K) moves the trolley towards the first display in the store, straight ahead. The display consists of punnets of strawberries.	
K: Oh look, here we go, I see strawberries. We need some, don't we?	*Designing:* expressing a symbolic plan.
Evie (E) is facing in a different direction, looking at the other displays. K turns and stops the trolley parallel to the first display of strawberries in punnets. E turns and looks at the sign on display beneath the punnets. K places two punnets of strawberries in the trolley.	*Locating:* changing position to orient herself towards the object in question. *Counting:* modelling the quantity 'two'.
E: *[looking at K]* Watermelon.	*Designing:* expressing a symbolic plan.
K: Watermelon, yes, we'll get watermelon. Did you see some? I didn't see it.	*Locating:* discussing the location of items.
E looks out into the store. E holds her gaze in one direction. K turns her head left and right quickly.	*Locating:* orienting herself towards the object in question.
E: Watermelon. *[E continues to look in the same direction].*	*Locating:* orienting herself towards the object in question.
K: Watermelon? *[K follows where E is looking]* Oh yes, I see some.	*Locating:* changing position to orient herself towards the object in question.
K moves trolley towards the watermelon, which E is still looking at.	*Locating:* navigating and positioning himself in the spatial environment.
E: Watermelon.	
K: Yes, here it is. Here's the watermelon. We want rockmelon, too. But I can't see that.	*Designing:* expressing the need for different categories of an item.
K picks up a watermelon and places it in the trolley.	
E: Rockmelon?	*Designing:* confirming the previously expressed plan for a different category of the item.
K: Yes, we want rockmelon. But first, Mummy wants to get some asparagus.	*Designing:* expressing a symbolic plan.
K moves the trolley away from the melons.	*Locating:* changing position to orient herself in a different direction.

Table 3. Vignette from Family 2.

Observations	Mathematical Activities
Corey (C): Should we get some more bananas?	*Designing:* expressing a symbolic plan. *Counting, measuring:* expressing quantity and comparisons of quantities.
Jasper (J), who had been facing in the opposite direction, turns to look at the bananas.	*Locating:* changing position to orient himself towards the object in question.
C: What do you reckon? Should we get two of them?	*Counting:* expressing a numerical quantity.
J: Yum yum.	
C puts two bananas in a bag and places them in the trolley.	*Counting:* modelling the quantity 'two'.
C: What else do we need?	
C looks down at the shopping list he is holding, and turns the trolley as he is looking down at the list. As he does so, the trolley comes very close to bumping into another shopper. J is watching as the trolley moves close to the shopper.	*Locating:* navigating and positioning himself in the spatial environment.
J: Oopsie!	*Locating:* noticing the spatial position of objects relative to one another.
C looks up from the list, and reverses the trolley. They continue moving along the aisle. J is looking at the different food items as they walk past.	*Locating:* navigating and positioning himself in the spatial environment.
J: No … no … [then] Daddy! Daddy! [J looks at the watermelon] Woohoo!	*Designing, measuring:* comparing different food items; seeking a specific item.
C turns to look where J is indicating.	*Locating:* changing position to orient himself towards the object in question.
C: Watermelon? Would you like to get a small one? [C picks up a watermelon]	*Measuring:* comparing sizes.
J: Yeah Daddy! Up it there!	*Designing:* expressing a symbolic plan.
C shows J the watermelon.	
C: Should we get this one?	
J: Up it there, up it there! [indicates that C should put the watermelon in the trolley].	*Locating:* indicating where the object should be placed.
C places the watermelon in the trolley.	*Locating:* placing the object in a specific location.

Nelson to find particular items, but the boys regularly check back with their father and engage in discussion throughout the shopping trip. In this vignette (Table 4), Brett, Nelson, Cash, and Rebecca are discussing which potatoes they should purchase. Damian is not present for this discussion.

Family 4

Family 4 mother, Sandie, is shopping with her three children: son, Zack (8 years); son, Callum (7 years); and daughter, Ava (3 years). Sandie is pushing a trolley, in which Ava is seated. Zack and Callum are walking beside her. In this vignette (Table 5), the family is walking down the meat aisle. Ava begins sharing her memories of a previous visit to this store.

Family 5

Family 5 mother, Jillian, is shopping with her daughter, Tilly (3 years) and son, Leo (12 months). Jillian is pushing a trolley; Leo is sitting in the child seat, facing Jillian, and Tilly is sitting in the trolley itself, facing forward. Tilly is holding the shopping list and a pen, and is marking off items as they shop. In this vignette (Table 6), the family is purchasing some sausages for dinner.

Family 6

Family 6 mother, Lucy, is shopping with her daughter, Isobel (6 years). Lucy is pushing a trolley and Isobel is walking beside her. Isobel has prepared a list of items which they are

Table 4. Vignette from Family 3.

Observations	Mathematical Activities
Brett (B): Hey Nelson, where's the potatoes, buddy?	*Locating:* identifying the position of the potatoes.
B pushes trolley towards potatoes section of store.	*Locating:* navigating the space to find a specific location.
Nelson (N): Red potatoes?	*Designing:* expressing desire for a particular category of potatoes.
B: Ahh … what's the best price?	*Measuring:* modelling the need to compare prices.
B pauses and looks at large price signs displayed above the different types of potatoes. Rebecca (R), seated in trolley, and Cash (C) standing beside the trolley, both follow B's gaze from the signs to the potatoes. R points to the red potatoes.	*Locating:* looking at the price signs. *Measuring:* comparing the prices. *Locating:* R responds to the previous request for red potatoes.
R: *[Quietly]* Red potatoes.	*Locating:* indicates that she has located the red potatoes.
B: Nine dollars … *[to R]* Hey?	*Counting:* expressing a numerical value.
R points to the red potatoes again.	*Locating:* indicates that she has located the red potatoes.
N: What about these ones? *[points to next type of potatoes]*	*Measuring:* comparing the different potatoes.
B: *[to N]* What's that one?	*Counting, comparing:* querying the prices of the different potatoes.
B, N and C all look at the pricing sign above the next type of potatoes. R. continues to look at the red potatoes.	*Locating:* looking at the price signs.
B: *[to N]* It's up there, look *[points to price sign]*. Both N and C look at the price.	*Locating:* directing attention to the price.
[Simultaneously] C: Seven dollars and ninety-nine cents. N: Seven-ninety-nine.	*Counting:* expressing the price of the potatoes.
R turns body to look where C and N are looking.	*Locating:* orients herself to the sign currently being discussed.
B: For five kilos … Yeah, that's pretty good.	*Counting:* expressing the value of the potatoes. *Explaining:* expressing that the potatoes are good value.
N: Is that ok?	
B: Yeah … But that's a lot of potatoes …	*Explaining:* confirming that the potatoes are good value. *Measuring:* comparing the quantities of potatoes.
B looks at the other types of potatoes.	
B: Hang on, these are little bit cheaper … they're tiny though, aren't they? They won't make very good chips.	*Measuring:* comparing the prices and sizes of the potatoes; modelling comparative language. *Explaining:* expressing why the size of the potatoes needs to be considered.
N moves over to where B is and looks at the potatoes. R and C also look at the potatoes. B points back at the potatoes previously being discussed.	*Measuring:* comparing the sizes of the potatoes.
B: *[to N]* Are those ones a bit bigger?	*Measuring:* modelling comparative language.
N looks back and forth between the potatoes. N touches the five kilo bag of potatoes previously being considered.	*Measuring:* comparing the sizes of the potatoes.
N: Yeah, these ones are ok.	*Explaining:* confirming that the size of the potatoes is suitable.

shopping for today. Isobel is carrying her list, and together she and Lucy are locating each of the items on Isobel's list. In this vignette (Table 7), Isobel is deciding what meat they should purchase.

Discussion

The results of this project suggest that young children explore a range of mathematical ideas whilst shopping with their family. In this section, we consider each of three research questions in turn and draw links to the body of background literature which has informed this study.

Table 5. Vignette from Family 4.

Observations	Mathematical Activities
Ava (A): I remember being in this shop. Yeah, when I was a baby. A tiny baby.	*Playing:* expressing a recreation of reality. *Measuring:* expression of size and age.
Sandie (S): A tiny baby?	
A: Yeah, I come here. And you let me have the food *[gestures towards the meat]* and I dropped it.	*Explaining:* providing details to confirm the expression of reality.
S: You dropped it?	
A: Yeah. And you had to pick it up off the floor.	*Locating:* describing the location of the object in question.
S: I don't remember that, Ava.	
A: Well, I did.	
A points towards the bottles of sauce on the other side of the aisle.	*Locating:* indicating a specific object.
A: That's not heavy things. *[A then looks back to the other side of the aisle and points to the meat].* But those things are heavy and I dropped them. *[A looks back at the sauce].* But not heavy things, I didn't drop not heavy things. When I was a tiny baby.	*Measuring:* using the language of mass measurement; comparing the masses of objects. *Locating:* changing spatial position to indicate objects being compared. *Measuring:* expression of size and age.

What mathematics can be noticed in family shopping experiences?

As outlined earlier, Bishop's (1988) six categories of mathematical activities are a representation of the mathematical actions demonstrated in a range of cultural contexts. All of Bishop's activities were evident within the context of shopping, with even the youngest children in this study displaying actions consistent with Bishop's activities. Among the youngest children (12 months–3 years), the activity of *locating* was most frequently observed, with children using gesture, gaze, speech, and speech utterances to indicate the location of things they had noticed, the position of specific items, and the position of items in relation to one another. *Designing* was also evident among these young children, as they expressed a desire for a particular item, or to move in a particular direction. Additionally, *playing* and *measuring* was evident in the vignette of 3-year-old Ava as she recreated her earlier experiences and the implications of picking up 'heavy things'.

Table 6. Vignette from Family 5.

Observations	Mathematical Activities
Jillian (J): What are you going to have for dinner?	
Tilly (T): Sausages?	*Designing:* expressing a symbolic plan.
J: Sausages! Ok, we'll come and get some sausages.	
J starts to move the trolley.	*Locating:* changing position to orient herself towards the object in question.
J: Can you see the sausages?	*Locating:* directs attention to the object in question.
T looks around and points to the deli counter.	*Locating:* indicates that the object in question has been located.
T: There it is!	*Locating:* indicates that the object in question has been located.
J: There they are. *[To the deli assistant]* Can I just grab half a kilo of those sausages down the back …	*Measuring:* modelling the language of quantity; expressing a specific quantity.
Leo (L) points to the sausages.	
L: Na-na! Na-na!	*Explaining:* indicates a familiar shape.
J: They're sausages. Not bananas – sausages!	*Explaining:* explains the category of item.
T: It's not na-nas, it's sausages.	*Explaining:* explains the category of item.
J places the sausages in the trolley.	
J: Ok, we've got the sausages, so we can cross them off the list.	
[J shows T how to cross the item off the list].	

Table 7. Vignette from Family 6.

Observations	Mathematical Activities
Lucy (L): Do you want to get some steak?	
Isobel (I) looks at her shopping list.	*Designing:* expressing a symbolic plan.
I: Yeah.	
L: There's steak here *[L points to the meat counter].*	*Locating:* indicating the location of the item in question.
I pauses and looks at the steak.	
L: There's T-bones …	*Explaining:* expressing different categories of the item in question.
I: No, little ones.	*Designing:* expresses the desire for a specific category of the item in question.
	Measuring: compares the sizes of the different categories of items; uses the language of size.
L: There's little ones over there *[L points to some eye-fillet steak].*	*Locating:* indicates the location of the desired item.
	Measuring: uses the language of size.
I looks back and forth between the steaks. She points to the eye-fillet steaks.	
I: The little ones.	*Designing:* confirms the desired size.
L: *[to the shop assistant]* We might get two of those *[points to the T-bone steaks]* and three eye-fillet steaks as well please.	*Counting:* uses the language of quantity.
	Designing: uses different categories of the item.
L: Ok, what else is on your list Isobel? We've got the meat and fish *[L points to the list]* … Some nuts.	
L and I continue moving through the store.	
L: Nuts … Let's have a look …	*Locating:* looking for the location of a specific item.
I: Can I get some salami?	*Designing:* expresses desire for another category of an item.
L: You want to get salami?	
I: Yep. That's meat, too.	*Explaining:* recognising different categories of an item.

The full range of Bishop's activities was evident among the older children (6–10 years), who engaged in a variety of mathematical processes as they shopped. *Locating* was again most prevalent, as children and parents discussed and located particular items. There were a number of instances of *counting* as families collected specific quantities of items, and parents modelled appropriate language to describe these quantities. *Measuring* was also evident as families compared quantities and attributes of items; in particular, the relative sizes of items (for example, potatoes, steak).

Explaining was observed among the data; and interestingly, in most instances, it was the children explaining the reasoning for their ideas or expressions. The vignette featuring Family 5 provides a particularly interesting example of *explaining*, as the 12-month-old infant connects the shape of a sausage to the shape of a banana – which appears to be a known word – and is corrected by his three-year-old sister. This example is indicative of the everyday, natural ways in which young children and their families interact about mathematics whilst shopping.

In what ways is it noticed, and by whom?

As demonstrated in the vignettes, all participants – children and adults – engaged in mathematical noticing, in different ways at different times. Several of the vignettes demonstrate how environmental print, in particular, can provide opportunities for noticing mathematics. For example, the vignette featuring Family 3 demonstrates how environmental print in the form of pricing signs can prompt a substantial discussion about cost and value, and provide opportunities for comparison. Consistent with Mason's (2002) framework, the participants demonstrated a variety of ways of 'marking' what they were noticing. The participants all used combinations of gesture, gaze, and speech (including speech utterances among the younger children) to 'mark' – or indeed, *remark* to one another – what they were noticing. The notion of non-verbal 'marking' is particularly evident in the vignette featuring Family 1, in which the young child consistently uses gesture, gaze, and bodily position to mark what she is noticing, and to communicate this noticing to her mother.

In what ways do children and families explore and talk about mathematics while shopping together?

All six families explored and talked about mathematical ideas in different ways, for different purposes. In some instances, the families *talked about* mathematics as expressions of 'intent'; for example, 'should we get two bananas?' (Family 2), and 'get the little steaks' (Family 6). The *measuring* activity of 'comparing' was often the basis for *exploring* and *talking about* quantities or qualities of items; for example, comparing both the cost and size of potatoes (Family 3), and talking about heavy and light food items (Family 4).

The vignette featuring Family 3 was an example of how families might explore and talk about mathematical concepts in flexible, but purposeful ways. The conversation between the father and children about which potatoes to purchase confirms Nunes, Schliemann, and Carraher's (1993) notion of 'street maths', in that the family simultaneously took into account the cost of the potatoes as well as their suitability for purpose ('they won't make very good chips') when assessing the relative value of the potatoes. These mathematical judgments occurred as a rapid series of rough comparisons and approximations; representative of 'everyday' uses of mathematics in meaningful contexts.

Conclusion

This pilot project has demonstrated that young children and their families notice, explore, and talk about a great deal of implicit and explicit mathematics whilst shopping together. All six of Bishop's mathematical activities were identified among the data. 'Locating' was the most frequently observed of Bishop's activities among the data, with many instances evident in relation to the families with the youngest children (12–23 months), in particular. All of the children displayed instances of mathematical noticing, with the children 'marking' what they had noticed in both verbal and non-verbal forms. Furthermore, all six families explored and talked about what was noticed whilst shopping together.

This project also confirmed that innovative visual methods such as 'trolley-cam' can facilitate researcher noticing of families' interactions around mathematics. The trolley-cam provided a new perspective on family shopping experiences and allowed the researchers to conduct a fine-grained analysis which made visible the mathematical activities and interactions which might otherwise have gone unnoticed. The use of Bishop's (1988) mathematical activities as an analytical framework prompted 'intentional noticing' (Mason 2002) by the researchers as they attended to the mathematical actions and interactions of the children and their parents.

This research has implications for school- or centre-based education because engagement in cultural practices outside of school may have a profound impact on the knowledge that children bring to the classroom (Guberman 2004). This study elucidates the mathematical activities and interactions in everyday contexts such as family shopping that contribute to children's ongoing mathematical learning and development outside of homes, early childhood centres, and schools. Early childhood educators must recognise and build upon children's everyday mathematics learning, and value children's everyday life experiences as rich resources for mathematical learning. This study contributes to a broader understanding of the foundations upon which early childhood mathematics education can be built.

Disclosure statement

No potential conflict of interest was reported by the authors.

Funding

This project was funded by an internal research grant from Charles Sturt University's Faculty of Arts and Education.

References

Anderson, A. 1998. "Parents as Partners: Supporting Children's Mathematical Learning Prior to School." *Teaching Children Mathematics* 4 (6): 331–337.

Aubrey, C. 1993. "An Investigation of the Mathematical Knowledge and Competencies Which Young Children Bring Into School." *British Educational Research Journal* 19 (1): 27–41.

Baroody, A. J. 2000. "Does Mathematics for Three and Four Year old Children Really Make Sense?" *Young Children* 55 (4): 61–67.

Bishop, A. J. 1988. *Mathematical Enculturation*. Dordrecht: Kluwer.

Björklund, C. 2008. "Toddlers' Opportunities to Learn Mathematics." *International Journal of Early Childhood* 40 (1): 81–95.

Clarke, B., D. Clarke, and J. Cheeseman. 2006. "The Mathematical Knowledge and Understanding Young Children Bring to School." *Mathematics Education Research Journal* 18 (1): 78–102.

Davidson, C. 2010. "Transcription Matters: Transcribing Talk and Interaction to Facilitate Conversation Analysis of the Taken-for-Granted in Young Children's Interactions." *Journal of Early Childhood Research* 8 (2): 115–131.

De Lange, J. 2008. *Talentenkracht* [Curious minds]. TFreudenthal Institute for Mathematics and Science Education. The Netherlands.

Duncan, G. J., C. J. Dowsett, A. Claessens, K. Magnuson, A. C. Huston, P. Klebanov, L. S. Pagani, et al. 2007. "School Readiness and Later Achievement." *Developmental Psychology* 43 (6): 1428–1446.

Fasoli, L. 2003. "Reading Photographs of Young Children: Looking at Practices." *Contemporary Issues in Early Childhood* 4 (1): 32–47.

Fenton, A. 2013. "Using a Strengths Approach to Early Childhood Teacher Preparation in Child Protection Using Work-Integrated Education." *Asia-Pacific Journal of Cooperative Education* 14 (3): 157–169.

Gervasoni, A., and B. Perry. 2015. "Children's Mathematical Knowledge Prior to Starting School and Implications for Transition." In *Mathematics and Transitions to School: International Perspectives*, edited by B. Perry, A. MacDonald, and A. Gervasoni, 47–64. Dordrecht: Springer.

Guberman, S. R. 2004. "A Comparative Study of Children's Out-of-School Activities and Arithmetic Achievement." *Journal for Research in Mathematics Education* 35 (2): 117–150.

Horner, S. L. 2005. "Categories of Environmental Print: All Logos Are Not Created Equal." *Early Childhood Education Journal* 33 (2): 113–119.

Joffe, H. 2011. "Thematic Analysis." In *Qualitative Research Methods in Mental Health and Psychotherapy: A Guide for Students and Practitioners*, edited by D. Harper and A. R. Thompson, 209–223. Chichester: John Wiley & Sons.

Klibanoff, R. 2006. "Preschool Children's Mathematical Knowledge: The Effect of Teacher 'Math Talk'." *Developmental Psychology* 42 (1): 59–69.

Linder, S. M., B. Powers-Costello, and D. A. Stegelin. 2011. "Mathematics in Early Childhood: Research-based Rationale and Practical Strategies." *Early Childhood Education Journal* 39: 29–37.

MacDonald, A. 2010. "Young Children's Measurement Knowledge: Understandings About Comparison at the Commencement of Schooling." In *Shaping the Future of Mathematics Education: Proceedings of the 33rd Annual Conference of the Mathematics Education Research Group of Australasia*, edited by L. Sparrow, B. Kissane, and C. Hurst, 375–382. Fremantle: MERGA.

MacDonald, A. 2012. "Young Children's Photographs of Measurement in the Home." *Early Years* 32 (1): 71–85.

MacDonald, A., and C. Carmichael. 2015. "A Snapshot of Young Children's Mathematical Competencies: Results from the Longitudinal Study of Australian Children." In *Mathematics Education in the Margins: Proceedings of the 38th Annual Conference of the Mathematics Education Research Group of Australasia*, edited by M. Marshman, V. Gieger, and A. Bennison, 381–388. Sunshine Coast: MERGA.

MacDonald, A., and C. Carmichael. 2016. "Early Mathematical Competencies and Later Outcomes: Insights from the Longitudinal Study of Australian Children." In *Opening up Mathematics Education Research: Proceedings of the 39th Annual Conference of the Mathematics Education Research Group of Australasia*, edited by B. White, M. Chinnappan, and S. Trenholm, 413–420. Adelaide: MERGA.

Macmillan, A. 2009. *Numeracy in Early Childhood: Shared Contexts for Teaching and Learning*. South Melbourne: Oxford University Press.

Mason, J. 2002. *Researching Your own Practice: The Discipline of Noticing*. London: RoutledgeFalmer.

McCashen, W. 2005. *The Strengths Approach*. Bendigo, VIC: St. Luke's Innovative Resources.

McNeill, D. 1992. *Hand and Mind: What Gestures Reveal About Thought*. Chicago: The University of Chicago Press.

Neumann, M. M., M. Hood, R. M. Ford, and D. Neumann. 2011. "The Role of Environmental Print in Emergent Literacy." *Journal of Early Childhood Literacy* 12 (3): 231–258.

Nunes, T., A. Schliemann, and D. Carraher. 1993. *Street Mathematics and School Mathematics*. Cambridge: Cambridge University Press.

Perry, B., and A. Gervasoni. 2012. *Let's Count Educators' Handbook*. Sydney: The Smith Family.

Reikerås, E., I. K. Løge, and A. Knivsberg. 2012. "The Mathematical Competencies of Toddlers Expressed in Their Play and Daily Life Activities in Norwegian Kindergartens." *International Journal of Early Childhood* 44: 91–114.

Saleebey, D. 2009. *The Strengths Perspective in Social Work Practice*. Boston: Allyn and Bacon.

Sarama, J., and D. H. Clements. 2015. "Scaling up Early Childhood Mathematics Interventions: Transitioning with Trajectories and Technologies." In *Mathematics and Transition to School: International Perspectives*, edited by B. Perry, A. MacDonald, and A. Gervasoni, 153–169. Dordrecht: Springer.

Tunnicliffe, S. 1995. "Talking About Animals: Studies of Young Children Visiting Zoos, a Museum and a Farm." PhD diss., King's College, University of London.

Watts, T., G. Duncan, R. Siegler, and P. Davis-Kean. 2014. "What's Past Is Prologue: Relations Between Early Mathematics Knowledge and High School Achievement." *Educational Researcher* 43 (7): 352–360.

Worthington, M., and E. Carruthers. 2003. *Children's Mathematics: Making Marks, Making Meaning*. London: Paul Chapman.

Wright, R. J. 1994. "A Study of the Numerical Development of 5-Year-Olds and 6-Year-Olds." *Educational Studies in Mathematics* 26: 25–44.

3   Mathematizing in preschool

Children's participation in geometrical discourse

Gabriella Gejard and Helen Melander ⓘD

ABSTRACT

This study explores preschool children's mathematizing in everyday block play activities. Building on an ethnomethodological and multimodal conversation analytic framework, we explore how geometry (i.e. spatiality, shape, and symmetry) is actualized in children's verbal and embodied interaction with their peers, pedagogues, and material environment. The selected data are drawn from a video ethnographic study in a Swedish preschool in which a boy and a girl play with a magnetic construction toy. The results of the study demonstrate how the participants orient to spatial locations, properties, dimensions, orientations, transformations, and shapes as they build a house. The children are shown to rely upon verbal and embodied resources such as deictics (e.g. here, there, these) and pointing gestures as geometrical aspects are actualized in their interaction. The study contributes with knowledge on preschool children's everyday mathematizing, in particular, children's appropriation of geometric discourse as it emerges in the unfolding flow of interaction.

Introduction

This study explores children's mathematizing in the block play area, with a particular interest in preschoolers' understandings of spatiality, shape, and symmetry, that is, geometry (see NCTM 2000; Seo 2003). Research about early childhood mathematics is a comprehensive field where children's counting, sorting, numeracy, and development of number sense are often focused (e.g. Baroody, Lai, and Mix 2006; Doverborg and Samuelsson 2000; Edens and Potter 2013). As yet, there are only a few studies that focus on children's development of understandings of geometry (although see e.g. Bäckman 2015; Casey et al. 2008), although there is a growing interest in how block play can contribute to children's early mathematical development (e.g. Albinsson 2016; Ness and Farenga 2007; Ramani et al. 2014). Previous research suggests that block play is of particular importance to children's development of spatial skills and understanding (e.g. Jirout and Newcombe 2015; Trawick-Smith et al. 2016).

To examine how children actively participate in geometrical discourse in everyday interaction in the block play area, we use an ethnomethodological and multimodal conversation analytic perspective (e.g. Goodwin 2000; Goodwin and Goodwin 2004). We analyze participation in mathematical discourse, *mathematizing*, as a socially organized process (Sfard 2008), based on video recorded data in which preschool children (aged 5 years) participate together with a pedagogue[1] in a block play activity using a magnetic construction toy in a Swedish preschool. In the analysis, we explore how the participants mobilize a multitude of multimodal resources such as talk, pointing, gesture, body orientation, and material structure in the environment as they make relevant geometrical aspects.

The study thus aims to contribute to research on children's everyday interactions during block play and its relation to development of mathematical understandings, in particular, geometry (cf. Clements 2001; Ness and Farenga 2007).

Research on geometry in block building play

Geometry is a mathematical field concerned with questions of shape, size, the relative position of figures, and the properties of space. Previous research has shown that playing with blocks promotes preschool children's development of understandings of geometry. For example, Casey et al. (2008) examined the effects of structured block building activities and found that block play encouraged children to test spatial relationships while building, in particular, when it was organized within the context of storytelling. The narrative motivated the children to build the structures the way the characters of the story requested, something that in turn contributed to make critical elements of the block building tasks more salient thus increasing the children's understanding of spatiality. Caldera et al. (1999) reported that preschoolers' block building skills appear to be related to their spatial visualization skills as measured by their ability to analyze and reproduce abstract patterns, to abstract a geometric figure embedded within a more complex figure, and to reproduce three-dimensional structures made from cubes. Ferrara et al. (2011) investigated if the context of block play had an impact on the amount of spatial language that children (aged 3–5) and their parents used in joint play sessions. They concluded that guided play contexts, in which the participants were given numbered photographs depicting the steps required to build a specific structure, elicited more spatial language both from parents and children compared with other play contexts (e.g. free play). Ramani et al. (2014) observed preschoolers aged 4–5 years building houses together with a peer in a guided block play activity. They found that children's spatial talk was intertwined with talk about features of the house that the children imagined they were building. The findings suggest that playing with peers may help children develop and expand their spatial understanding as they engage in spatial talk while building. In all, these studies indicate that there is a relationship between block building play and children's development of spatial understandings. However, they are all conducted with an experimental design in the shape of interventions and guided play activities, sometimes accompanied by pre- and post-tests of children's spatial skills and understandings.

In contrast, some studies argue the importance of studying children's everyday interactions. Based on case studies, Ness and Farenga (2007) show how children develop geometric, spatial, and scientific skills in free play with blocks. They found that when the children engaged in block play activities, spatial, and geometric concepts as well as architectural principles were actualized and that playing with blocks had a positive impact on children's mathematical behavior in general. Through observations of children's free play, Clements (2001) found that children deal with horizontal lines, parallelism, and symmetry while building with blocks. Important conclusions from studies such as these are that they show the complexity of everyday mathematics where there is an interplay between social interaction and material environment and where interaction with peers drives the children's explorations of geometrical aspects (see also Trawick-Smith et al. 2016).

Studies investigating preschool children's use of gestures and other embodied resources to express geometrical understandings are rare. However, analyzing kindergartners' use of gestures while playing a game in which the children were asked to describe geometrical shapes to their peers, Skoumpourdi (2016) found that the children used gestures independently of talk, producing iconic gestures that mimicked (more or less successfully) the shape that they were asked to describe. Elia, Gagatsis, and van den Heuvel-Panhuizen (2014) studied a five-year-old's learning of spatial and shape concepts in interaction with a teacher in which the child was asked to describe different spatial arrangements of blocks. The results of the study show the role of gestures in using and communicating spatial and shape-related ideas, in particular, deictic and iconic gestures and provide further evidence for the strong interrelations between geometrical thinking and gestures.

Studies of everyday block play activities are still few, and more research is needed (Ness and Farenga 2007). In such a vein, the present study aims to contribute to this emerging field of research with knowledge about how geometrical aspects such as spatiality, shape, and symmetry are actualized in children's verbal and embodied interaction with their peers, pedagogues, and material environment in everyday block play activities.

Mathematizing in interaction

The concept of mathematizing has recently received much attention in mathematics education research (e.g. Reis 2011; Sarama and Clements 2009; Sfard 2008; van Oers 2014). Sfard (2008) defines *mathematizing* as participation in mathematical discourse. Mathematical discourse is a discourse about mathematical objects, where mathematics emerges as a system that 'contains the objects of talk along with the talk itself and grows incessantly "from inside" when new objects are added one after another' (Sfard 2008, 129). Thus, mathematical objects do not pre-exist talk about them but evolve in and through talk. A mathematical discourse involves word use (e.g. words that signify quantities and shapes), visual mediators (visible objects or symbolic artifacts), narratives (e.g. descriptions about relations between objects), and routines (repetitive patterns that are characteristic of a given discourse) (Sfard 2008, 133–134).

We use the notion of mathematization to highlight children's participation in geometrical discourse, emphasizing its collective and colloquial (i.e. everyday or spontaneous) characteristics. Building on an ethnomethodological and multimodal conversation analytic approach (EM/CA) (e.g. Goodwin 2000), we analyze mathematizing as a socially organized process by examining the methods participants use to accomplish social actions in naturally occurring interaction. The concept of *participation* foregrounds the interactive work of both speakers and hearers and provides a framework for exploring 'how multiple parties build action together while both attending to, and helping to construct, relevant action and context' (Goodwin and Goodwin 2004, 240). Participants in social interaction mobilize a set of multimodal resources (talk, embodied action, and material environment) for the locally situated, intersubjective, and methodic organization of interaction. The analyses thus trace how social action gradually evolves; how participants produce intelligible and accountable actions whilst interpreting and acting upon publicly displayed and mutually available actions. As we analyze children's participation and meaning-making in a block play activity, we explore how young children appropriate a geometric discourse. Children become active participants through their participation in collective activities within cultural settings in which they display their understandings of forms of culturally based knowledge (cf. Aarsand and Melander 2016; Martin and Evaldsson 2012), in our case, understandings of spatiality, shape, and symmetry.

Research method

Video ethnographic fieldwork, empirical setting, and analyzed activity

The analyses are based on video recordings from a seven-month long video ethnographic study documenting mathematical activities in a Swedish preschool (Gejard 2014). The pedagogues actively planned for mathematical explorations as part of everyday life, which was an important reason for conducting fieldwork at this location. The data that will be analyzed here consist of an appr. 22 min. long recording of interaction in the block play area, in which two children, a girl (5.3 years) and a boy (5.2 years), whom we are calling Hanna and Elias, are participating together with one of the pedagogues, Lisa. Hanna and Elias recurrently participated in geometrical activities organized by the pedagogues, such as reading books and talking about geometrical shapes, making shapes from play dough, singing about shapes, etc. thus familiarizing themselves with basic geometric shapes and concepts. The block play area was located in one corner of a large room and it was furnished with a low table with a mirror on top. Along the walls, there were low shelves with building materials that aimed to encourage the children's mathematical explorations. The recorded activity took place before lunch, during free play, when the children were playing with a magnetic construction toy called Geomag. The toy consists of magnetic rods and metal spheres that can be attached to each other (Figure 1). There are plastic panels in the shape of squares, pentagons, and triangles that can be inserted into the shapes created by the rods and spheres.

Analytic approach and ethical considerations

With an interest in how participation is organized within the block play activity, and how the participants orient to spatiality, shape, and symmetry, the analysis follows the unfolding organization of one activity. Detailed attention is paid to how geometry is made relevant in children's interaction with their peers, pedagogues, and the material environment by analyzing the verbal and embodied resources (including talk, gestures, gaze, body positioning, body movements, and object manipulation) that participants draw upon to construct social action (e.g. Goodwin 2000). The children recurrently participated in block building play and the activity chosen for analysis is a typical example of what the children do in the blockplay

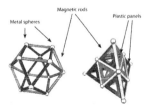

Figure 1. The Geomag magnetic construction toy.

corner and the geometry that is actualized during these activities. The excerpts selected for analysis contain references to lexical items that indicate spatial categories: spatial location (e.g. up, down etc.), spatial dimension (e.g. long, high etc.), spatial features or properties (e.g. curvy, straight etc.), shape (e.g. rectangle, square etc.), spatial orientation or transformation (e.g. turn it around etc.), and deictic terms (e.g. here, there etc.) (cf. Ferrara et al. 2011). The excerpts have been transcribed using conversation analytic conventions (see Appendix). In order to highlight embodied actions and the participants' orientations to the material environment, drawings based on frame grabs from the videos are included in the excerpts.

Written consent has been secured from the participating pedagogues as well as the children's guardians. The children were informed about the study and their right to choose whether they wanted to participate in the recordings or not during fieldwork. All participants have been given pseudonyms in order to protect their identities.

Mathematizing in the block play area

The analysis traces the evolving organization of a block play activity and we will show how the children make relevant a geometrical discourse while building. The first part of the analysis (Excerpt 1) demonstrates how the children orient to a number of spatial categories as they work with their constructions. In the second part of the analysis (Excerpt 2), one of the children asks the pedagogue a question that initiates a sequence in which they collaboratively explore geometrical shapes and spatial properties. Then follows a section in which spatial orientation and transformation are actualized as the children have decided to join their two constructions into one (Excerpt 3). Finally, the last part of the analysis (Excerpt 4) shows how symmetry plays a significant role as the children negotiate the design of the joint construction. Throughout the activity, Elias is sitting on the table and Hanna is on her knees on the floor opposite Elias. The pedagogue watches over the children, sometimes commenting on the constructions or responding to the children's questions.

Making relevant different spatial categories

We will begin by showing how the children make relevant spatial relationships. The excerpt is from the beginning of the activity, and the children are working on two different constructions. However, they actively engage in each other's projects by commenting upon them. In the first part of the excerpt, the children show each other their constructions, highlighting something 'worth seeing', whereas in lines 12–16, the two constructions are compared to each other.

In line 1, Elias calls for Hanna's attention with an instruction to look into his construction as he points at it, thus establishing a joint focus of attention. Hanna bends down, looks from the side, and responds by producing a response cry *wo̞:w* (Goffman 1978) followed by a sound object *eeh*, thereby showing appreciation of his structure (fig. 1.1). Elias elaborates on the feature that Hanna should observe: *it looks like a- (it's a) hole* (line 5). Rather than responding to the description of the construction in terms of a hole, Hanna produces a similar attention-seeking utterance: *look, now it looks like there are three: on mine* as she points at the bars (fig. 1.2). The utterance reuses resources provided by Elias' prior action to highlight an aspect of Hanna's construction, a way of building new action by performing operations on existing actions, something that has been shown to be central to

Excerpt (1): Look into mine
[22-130305; 00.40-02.10]

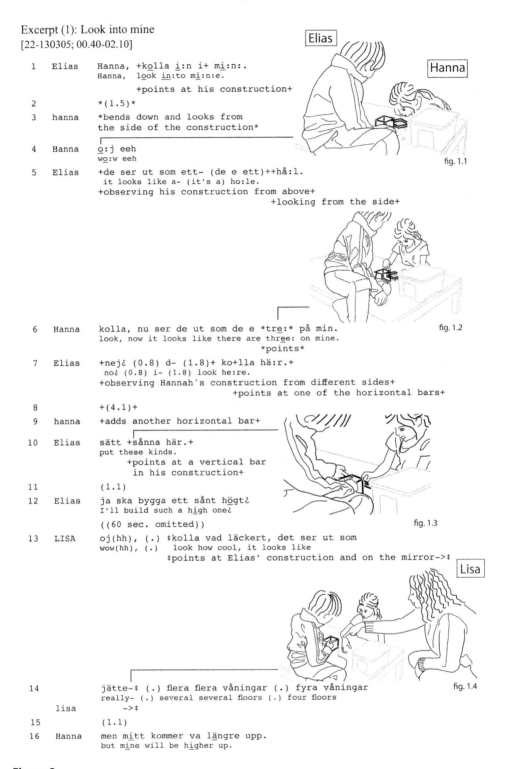

```
 1   Elias    Hanna, +kolla i:n i+ mi:n:.
                  Hanna,  look in:to mi:n:e.
                           +points at his construction+
```
Elias

Hanna

```
 2            *(1.5)*
 3   hanna    *bends down and looks from
              the side of the construction*

 4   Hanna    o:j eeh
              wo:w eeh
```
fig. 1.1

```
 5   Elias    +de ser ut som ett- (de e ett)++hå:l.
              it looks like a- (it's a) ho:le.
              +observing his construction from above+
                            +looking from the side+
```

fig. 1.2

```
 6   Hanna    kolla, nu ser de ut som de e *tre:* på min.
              look, now it looks like there are three: on mine.
                              *points*

 7   Elias    +nej¿ (0.8) d- (1.8)+ ko+lla hä:r.+
              no¿ (0.8) i- (1.8) look he:re.
              +observing Hannah's construction from different sides+
                            +points at one of the horizontal bars+

 8            +(4.1)+
 9   hanna    +adds another horizontal bar+

10   Elias    sätt +sånna här.+
              put these kinds.
                     +points at a vertical bar
                      in his construction+

11            (1.1)
12   Elias    ja ska bygga ett sånt högt¿
              I'll build such a high one¿

              ((60 sec. omitted))
```
fig. 1.3

```
13   LISA     oj(hh), (.) ‡kolla vad läckert, det ser ut som
              wow(hh), (.)  look how cool, it looks like
                            ‡points at Elias' construction and on the mirror->‡
```
Lisa

```
14            jätte-‡ (.) flera flera våningar (.) fyra våningar
              really- (.) several several floors (.) four floors
     lisa        ->‡
```
fig. 1.4

```
15            (1.1)
16   Hanna    men mitt kommer va längre upp.
              but mine will be higher up.
```

Figure 2.

how participants grasp the meaningfulness of the ongoing interaction (Goodwin 2013). Observing the construction from different sides, Elias rejects Hanna's observation: *no ¿(0.8) i- (1.8) look he:re* while pointing at one of the horizontal bars of her construction, possibly orienting to the fact that from his perspective the construction has four levels (2 in the construction +2 that are reflected in the mirror). Hanna continues building without attending to Elias. He does not further elaborate but watches her, and then suggests that she put additional bars in her construction by saying *put these kinds.* (line 10), pointing with an open hand to his own construction (fig. 1.3), emphasizing its height by moving his hand along the vertical bar. When no uptake is forthcoming from Hanna, Elias picks up some bars and declares *I'll build such a high one¿* as he continues his own project.

After a minute when the children have continued to build (not in the transcript), the pedagogue makes relevant the height of Elias' construction. Pointing at the construction and the mirror, she reorients to what Elias said in line 12 by producing a response cry followed by a positive assessment – *wow, (.) look how cool,*. She moves her pointing finger up and down along the side of the construction as she twice repeats 'several' and points down into the mirror, thus emphasizing its height and how it is reflected in the mirror, specifying the height of the building as *four floors*, referring to how the building appears in the mirror (2 + 2) (fig. 1.4). After a brief silence, it is Hanna who responds by declaring that hers will be higher: *but mine will be higher up.* The utterance is responsive to the pedagogue's positive assessment of Elias' work, where Hanna compares her future construction with the one that has just been assessed.

When building with the construction toy, different spatial categories are actualized. When Elias tells Hanna to look 'into' his construction, this indicates direction and is related to spatial locations. By referring to a 'hole' (line 5), the three-dimensional aspect of the construction, space, is highlighted. The children also use words that refer to spatial dimensions, such as when Elias declares that he will build 'such a high one' (line 12; cf. Ferrara et al. 2011). This can be contrasted to Hanna's 'higher up' (line 16) that is a location word; location words describing the position of an object, whereas dimension words describe the size of an object relative another object (Ferrara et al. 2011). The participants use *environmentally coupled gestures* (Goodwin 2007), that is, actions that rely upon talk, embodied action (pointing gestures), and the material environment, thus creating a powerful multimodal package of complementary meaning-making practices. Pointing gestures in combination with deictic terms are for example used by Elias to identify an object to attend to: 'look here' (line 7) and 'put these kinds' (line 10). The powerfulness of these combinations is displayed in how the children with rather limited verbal resources ('put these kinds') produce interactionally meaningful actions.

Exploring geometrical shapes and spatial properties

Between excerpts (1) and (2), the children have rebuilt their constructions. Hanna is now working on a cubic shape, similar to Elias' construction. In the analysis of excerpt (2), we will demonstrate how shapes and spatial properties are made relevant, as Hanna explores her construction by orienting to its straightness. The construction consists of magnetic rods and metallic spheres but no plastic panels, making the construction flexible and slightly unstable. Hanna holds her construction up in the air and asks the pedagogue if it is straight. Her question initiates a sequence in which Lisa and Hanna explore shapes and their properties.

Excerpt (2): Is mine straight?
[23-130305; 02.03-02.32]

```
   1   Hanna    e min ra:*+k?
                is mine strai:ght?
                          *holds her construction in the air, showing Lisa->*

   2            (1.5)
```

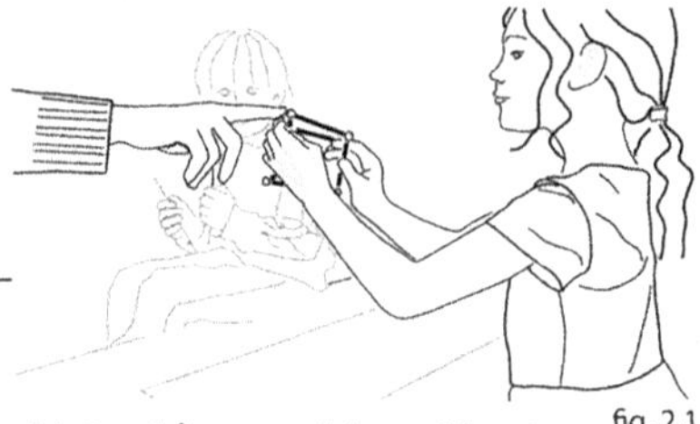

fig. 2.1

```
   3   LISA     ja:¿ d- (1.0) ‡titta. de e *li:ka långa‡ sider alla.*
                yeah¿ i- (1.0) look. all the sides are e:qually long .
                              ‡points at one of the sides‡

   4   hanna                        ->*drops the construction on the mirror
                                       so that it becomes distorted*
```

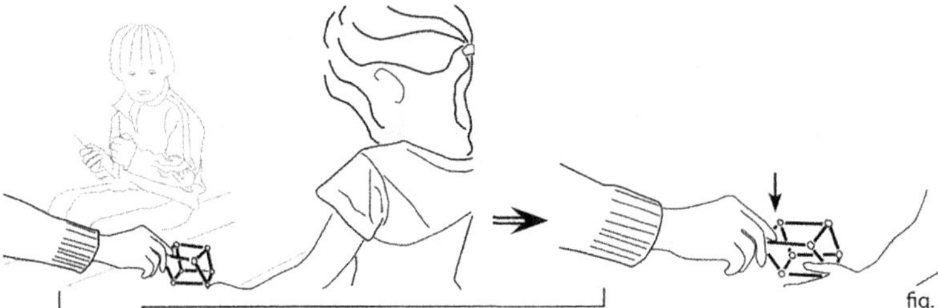

fig. 2.2

```
   5   LISA     >‡fast nu-‡
                but now-
                ‡presses one of the sides‡

   6            *(2.0)

   7   hanna    *straightens the construction so that it retains its cubic shape->*

   8   LISA     e den inte riktigt ra:k¿ nu blir de som en=
                isn't it really strai:ght¿ now it'll be like

   9   LISA     =annan [fo:rm.]  (.) än en kvadrat- (.) kub,
                another sha:pe.  (.) than a square- cube,

  10   Hanna          [mm*    ]
                       ->*
```

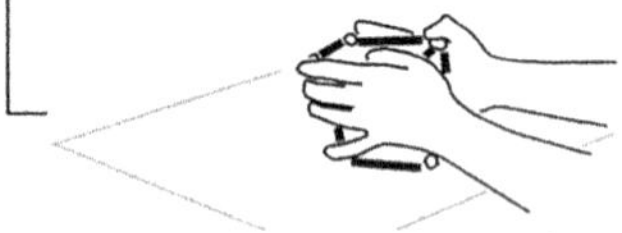

fig. 2.3

```
  11   Hanna    jag måste använda *sånna [*hä:r¿*
                I've got to use some of the:se¿
                                  *takes a square panel from the box*

  12   Elias                         [+tre plus tre blir sex.+ (1.0)
                                       three plus three is six.
                                       +holds bars in the air, showing Lisa+

  13   LISA     ja: ↑de blir e.
                yea:h ↑it is.

  14   Hanna    *de e de bästa du ve:t¿ ((singing))
                it's the best you kno:w¿
                *attaches a panel to the construction->*

  15   Hanna    nu kommer den inte att *va sne:.*
                now it won't be twis:ted.
                          ->*straightens the shape*
```

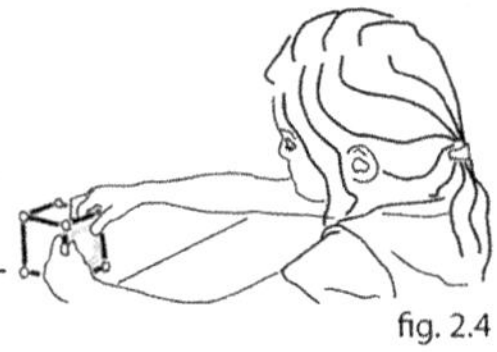

fig. 2.4

Figure 3.

By asking the pedagogue *is mine strai:ght?* (line 1), Hanna is making relevant spatial properties of her construction. The pedagogue hesitatingly confirms, possibly because of the fact that the construction has become slightly twisted as Hanna is holding it in her hands (fig. 2.1) and initiates a description of the construction *yea:h¿ i- (1.0) look. all the sides are e:qually long.* (line 3) as she points along one of the side bars. The question of straightness is responded to as being about the equally long sides. Hanna drops the construction on the mirror, distorting its shape. The pedagogue uses this as an educational resource, and she points and presses at the construction so that it becomes oblique, acquiring a rhombic shape (fig. 2.2). She comments on the shape of the construction, describing it as not really straight (line 8) and as being of a different shape (line 9). During this utterance, Hanna has straightened the construction (fig. 2.3), and Lisa continues by elaborating on the type of shape *than a square-cube,* (line 9). The pedagogue thus introduces a geometric discourse by first describing the construction as having equally long sides – the defining criteria for the geometric shapes square and cube. The notions are introduced as she denominates the object as having a different shape than a square, which is repaired by replacing *a square* with *a cube*. The former is not incorrect as a cube consists of six identical squares, but by using the term cube she makes relevant the three-dimensional shape of the construction. Hanna, who is still oriented to the spatial properties of the object (straightness), picks up a square panel, attaching it to the construction so that it acquires a straight shape: *now it won't be twis:ted* (line 15, fig. 2.4).

In excerpt (2), the participants use the flexibility of the magnetic construction toy to explore geometrical aspects. Hanna has been demonstrated to mainly orient to spatial properties of the object under construction, whereas the pedagogue oriented to the geometrical shapes (square, cube, equally long sides) that emerge as the construction is manipulated and transformed (cf. Ferrara et al. 2011).

Explaining spatial transformation and orientation

We will now show how geometrical aspects connected to spatial transformation and orientation are made relevant as the children decide to join their two constructions into one. The constructions are identical and in the shape of two cubic houses with a roof and a door. In similarity to how the pedagogue in excerpt (1) talked about Elias' construction in terms of having several floors (line 14), the children now talk about their constructions as representations.

Excerpt (3): You can turn it like this
[23-130305; 08.27-09.01]

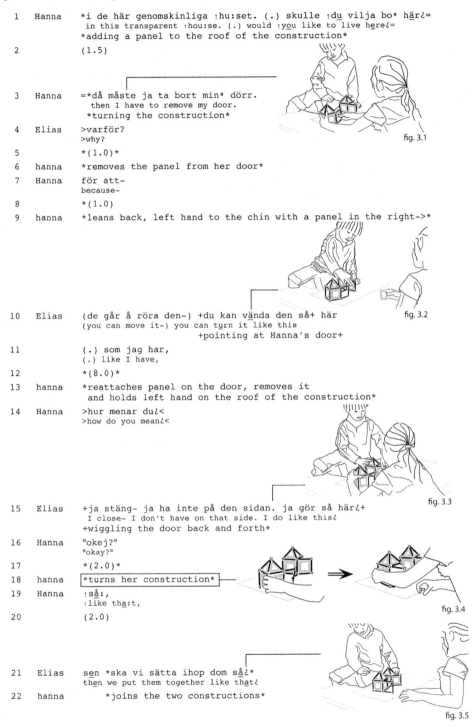

```
 1   Hanna    *i de här genomskinliga ↑hu:set. (.) skulle ↑du vilja bo* här¿=
              in this transparent ↑hou:se. (.) would ↑you like to live here¿=
              *adding a panel to the roof of the construction*

 2            (1.5)


 3   Hanna    =*då måste ja ta bort min* dörr.
              then I have to remove my door.
              *turning the construction*

 4   Elias    >varför?
              >why?

 5            *(1.0)*
 6   hanna    *removes the panel from her door*
 7   Hanna    för att-
              because-

 8            *(1.0)
 9   hanna    *leans back, left hand to the chin with a panel in the right->*




10   Elias    (de går å röra den-) +du kan vända den så+ här
              (you can move it-) you can turn it like this
                                 +pointing at Hanna's door+

11            (.) som jag har,
              (.) like I have,

12            *(8.0)*
13   hanna    *reattaches panel on the door, removes it
               and holds left hand on the roof of the construction*

14   Hanna    >hur menar du¿<
              >how do you mean¿<




15   Elias    +ja stäng- ja ha inte på den sidan. ja gör så här¿+
              I close- I don't have on that side. I do like this¿
              +wiggling the door back and forth+

16   Hanna    °okej?°
              °okay?°

17            *(2.0)*
18   hanna    *turns her construction*
19   Hanna    ↑så:,
              ↑like tha:t,

20            (2.0)




21   Elias    sen *ska vi sätta ihop dom så¿*
              then we put them together like that¿

22   hanna       *joins the two constructions*
```

fig. 3.1

fig. 3.2

fig. 3.3

fig. 3.4

fig. 3.5

Figure 4.

Hanna frames the activity as a playful event when she asks Elias *in this transparent ↑hou: se (.) would ↑you like to live here¿* (line 1). After a short silence during which Hanna and Elias both gaze at Hanna's construction, Hanna says that she needs to remove her door (line 3), thus identifying the door as an obstacle to the joining of the two constructions. Elias immediately asks Hanna why. Hanna removes the plastic panel from the door and then initiates an answer (*because-*) that is abandoned, thus displaying her difficulty in providing an account for why the door should be removed. Hanna's construction is placed on the mirror with the door directed toward Elias' construction, whereas Elias' construction is positioned with the door facing Hanna (fig. 3.1). Hanna thus seems to be oriented toward joining the constructions on her door side but displays uncertainty by leaning back and moving her hand to the chin, observing the constructions from some distance.

Elias produces an instruction (*you can move it-*) *you can turn it like this (.) like I have* (lines 10–11), touching the door on Hanna's construction as the first part of the utterance is produced, then briefly pointing at his construction, tracing the direction of the door (fig. 3.2). He thus suggests that Hanna turn her construction in the same direction as his. An 8 sec. silence follows, during which Hanna first reattaches and then removes the plastic panel on the door to then hold the construction with her hand on the roof, in all displaying that she is uncertain about what to do. A request for an explanation follows as Hanna asks > *how do you mean¿*< (line 14). Elias answers *I close- I don't have on that side. I do like this¿* as he wiggles the door, back and forth. The utterance is produced with several reformulations, displaying his difficulty to find words to explain what he means. He makes relevant direction ('that side') and describes a movement ('I do like this'). Hanna silently receipts Elias' explanations and then turns her construction, placing it in the same direction and with the same orientation as Elias' construction (fig. 3.4). The constructions are aligned and as Elias says *then we put them together like that¿* Hanna joins the two constructions (fig. 3.5).

The fact that the constructions are identical but placed in different directions poses problems to the children when joining the two constructions. It is a challenging task, but the children display a high level of engagement in the activity and a resolution to solve the problem, something that is visible in the way that they make suggestions, ask questions, and attempt to explain to each other what to do and why. The complexity of the task is shown by the children's difficulties in providing verbal explanations and where they instead rely upon pointing gestures, deictics, and embodied demonstrations ('you can turn it like this', 'I do like this') to indicate what they are referring to (e.g. lines 10, 15) (cf. Elia, Gagatsis, and van den Heuvel-Panhuizen 2014). The problem that they are facing has to do with direction, spatial orientation, and transformation (mental rotation, Casey et al. 2008). The way Elias' talk is produced with restarts and reformulations (lines 10–11, 15) demonstrates the complexity of describing an orientation or an act of moving objects in space. In addition, Hanna displays difficulties in finding out in what direction to turn her construction, but once she finds out what to do she rather quickly rotates it so that it ends up in the same direction as Elias', an action that can be understood as a precursor to mental rotation.

Negotiating symmetries

In this last section, we will highlight how the children make relevant symmetries as they negotiate the design of their construction. Between excerpts (3) and (4), the children have been working on their joint construction. The positions of the doors are the same, but they have joined the two doors with vertical bars (fig. 4.1). Hanna, who focused on the position of her door in excerpt (3), now argues that they have to change the position of Elias' door.

Excerpt (4): You've got to have your door there
[23-130305; 10.13-10.48]

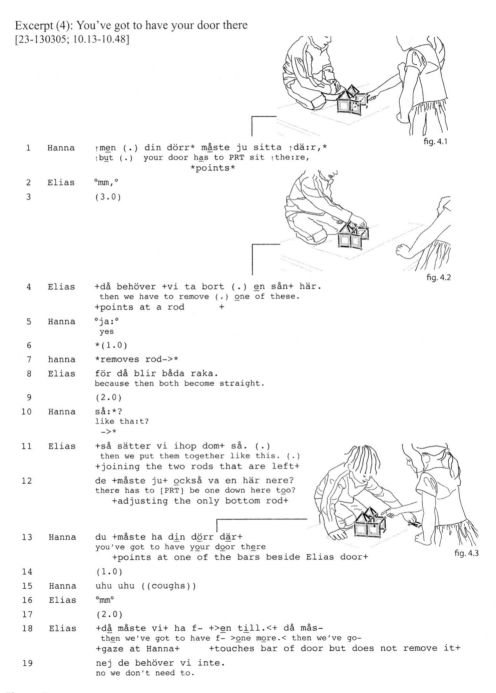

```
 1   Hanna     ↑men (.) din dörr* måste ju sitta ↑dä:r,*
                ↑but (.)  your door has to PRT sit ↑the:re,
                              *points*
 2   Elias     °mm,°
 3             (3.0)
```
fig. 4.1

fig. 4.2

```
 4   Elias     +då behöver +vi ta bort (.) en sån+ här.
                then we have to remove (.) one of these.
                +points at a rod       +
 5   Hanna     °ja:°
                yes
 6             *(1.0)
 7   hanna     *removes rod->*
 8   Elias     för då blir båda raka.
                because then both become straight.
 9             (2.0)
10   Hanna     så:*?
                like tha:t?
                 ->*
11   Elias     +så sätter vi ihop dom+ så. (.)
                then we put them together like this. (.)
                +joining the two rods that are left+
12             de +måste ju+ också va en här nere?
                there has to [PRT] be one down here too?
                    +adjusting the only bottom rod+
13   Hanna     du +måste ha din dörr där+
                you've got to have your door there
                    +points at one of the bars beside Elias door+
14             (1.0)
15   Hanna     uhu uhu ((coughs))
16   Elias     °mm°
17             (2.0)
18   Elias     +då måste vi+ ha f- +>en till.<+ då mås-
                then we've got to have f- >one more.< then we've go-
                +gaze at Hanna+      +touches bar of door but does not remove it+
19             nej de behöver vi inte.
                no we don't need to.
```
fig. 4.3

Figure 5.

Hanna orients to the problem that she perceives with Elias' door, which is that his door should be positioned further out on the construction. She points at the lower right end, indicating the correct position of the door (fig. 4.1) and saying ↑but (.) your door has to [PRT] *sit* ↑*the:re* (line 1). Elias minimally acknowledges, but rather than immediately responding to the position of the door, suggests that they remove a rod as he points at one of the bars between the two doors (fig. 4.2). Hanna, answers *yes* in a quiet voice, and then begins to remove one rod. Although the action is already on its way, Elias produces an account at this time, explaining why it was necessary to remove the rod: *because then both become straight.* (line 8), an utterance that refers to the two doors that will be straight in relation to each other. That Elias orients to straightness is reinforced as he helps Hanna adjust the rods and comments on the collaborative construction: *then we put them together like this. (.) there has [PRT] to be one down here too?* (lines 11–12). Adding a bar at the bottom of the construction increases its stability. The Swedish epistemic adverb 'ju' appeals to shared knowledge, meaning that the claim that there has to be a bar at the bottom is produced as something that both Elias and Hanna (should) know (Heinemann, Lindström, and Steensig 2011). However, before either child has added a bar, Hanna, who has actively engaged in Elias' project of straightening the construction, returns to the question of the position of the door, *you've got to have your door there* as she points at the far right side of the construction (line 13, fig. 4.3). She emphasizes 'your', 'door' and 'there', in all underlining crucial parts of the directive. After a silence and a minimal acknowledgment Elias aligns with Hanna's directive and moves his hand to the door, trying to detach the door from the construction (line 18). However, he quickly abandons the attempt and instead declares *no we don't need to.* This is accounted for by Elias (not in the transcript) by suggesting that the far end of the construction is an entrance, something that Hanna agrees with and the discussion about the door position comes to a close.

While building, the children make relevant spatial categories such as spatial properties ('straight'), spatial location ('there'), and spatial position ('down here'). Deictics are frequent and combined with pointings they highlight relevant aspects of the construction. The use of deictics presupposes that the addressee is attending to a specific place on the construction (see Goodwin 2007). The way in which the children work to secure each other's attention, attempting to establish a shared understanding of the activity at hand, is crucial for how they manage to collaboratively build a complex construction.

As the children negotiate the design of their construction, they create and combine different shapes and as the construction gets more complex they have to figure out how to coordinate the parts into a connected whole (cf. Casey et al. 2008, 272). In this process, they make relevant symmetry. On the one hand, changing the door position the way Hanna wishes would create a symmetrical mirrored construction, where what Hanna is orienting to is plane symmetry (Ness and Farenga 2007). On the other, Elias orients to symmetry when he suggests that they remove a rod (line 4) so that the doors 'become straight' (line 8), or in other words that the rods become parallel to each other, thus reinforcing the straightness of the shape. This is an example of line symmetry, where an object is placed equidistant from the sides of a larger structure, creating a symmetric appearance (Ness and Farenga 2007).

Concluding discussion

Children's participation in incipient geometrical discourse

In this paper, we have explored mathematizing in a naturally occurring, everyday block building activity, with an interest in children's displayed understandings of geometry. Through the detailed analyses of the unfolding organization of the activity, grounding the analyses in the participants' orientations rather than in predefined categories, we have shown how the children orient to and actualize different geometrical aspects as they work on their constructions: spatial locations, properties, dimensions, orientations, transformations, and geometrical shapes. The results of the study show the richness of children's spontaneous mathematical interactions and the number of geometric aspects that arise in their interaction as they proceed to collaboratively solve the problems that they encounter (cf. Clements 2001; Ness and Farenga 2007).

Describing the characteristics of a mathematical discourse, Sfard (2008, 133) highlights the importance of the words that participants use, as they are responsible for 'what the user is able to say about (and thus to see in) the world'. When the children in our data interact in the block play activity, they use different linguistic resources where some directly refer to geometrical concepts, for example, 'is mine straight'. More importantly, however, the children rely on unspecified notions and deictics such as 'look into mine', 'I've got to use some of these', 'put these kinds'. Elia, Gagatsis, and van den Heuvel-Panhuizen (2014) demonstrate how a child spontaneously used iconic and deictic gestures through-out a construction activity in order to describe different geometric shapes and relation-ships. In a similar vein, Skoumpourdi (2016) shows how children use different communicative modes to describe shapes and how gestures were found to function inde-pendently of oral language. In contrast, the results of our study show the symbiotic relationship between talk, gesture, and material environment, where talk and gesture mutually elaborate upon each other as part of collaborative meaning-making practices (cf. Goodwin 2000, 2007; Sfard 2009). Together with pointings and gestures, that in a sim-ultaneous and coordinated way are used to indicate specific objects, directions, or places, the children produce meaningful and mutually intelligible actions that contribute to and are indeed a condition for how the activity and the construction develops over time, from a number of metallic spheres and rods that are combined into two cubic shapes that are later joined and transformed into a complex building. The physical manipulation of the objects are here part and parcel of the children's geometrical discourse (cf. Sfard 2008, 148), and the material environment, that is, the Geomag construction toy, the mirror as well as how the children are positioned around the table, all contribute to the exploration of spatial phenomena such as height or straightness. In other words, the results of our study demon-strate the intertwinedness of talk, embodied actions, and the material environment in chil-dren's mathematizing, aspects of children's displays of mathematical understandings that only to a very limited extent have been investigated in previous research on preschool mathematics and geometry.

The children in our data are clearly newcomers to mathematical discourse, but through their active participation in everyday block play activities such as the one analyzed here, we argue that they appropriate an incipient geometrical discourse as it emerges in the unfold-ing flow of interaction. However, the research reported here calls for more studies on chil-dren's development of understandings of geometry as they are actualized in everyday, naturally occurring interactions – both in spontaneous play activities and in instructional activities organized by pedagogues. Our study represents a very first step in this direction and more research is needed in order to fully grasp the relation between talk, embodied actions, and other semiotic resources and its relevance for young children's development of geometrical thinking.

The findings reported in this study can inform pedagogues about ways of thinking of geometrical discourse as embodied, situated, and in dialogue with a material environment (cf. Elia, Gagatsis, and van den Heuvel-Panhuizen 2014). Our findings underline the importance of striking a balance between, on the one hand, controlling the children's activities, providing them with correct concepts and explanations and, on the other hand, letting their actions drive the activity forward, allowing them to explore geometrical shapes and relations in unpredictable ways. The detailed analyses of multimodal inter-action in everyday activities may help pedagogues become aware of children's geometrical knowledge that goes beyond verbal articulations of geometric concepts, in order to identify learning needs and to design pedagogical challenges.

Note

1. We use the word 'pedagogue' to denominate staff working at the preschool, as this was the way the participants in the study referred to themselves.

Disclosure statement

No potential conflict of interest was reported by the authors.

ORCID

Helen Melander ⓘ http://orcid.org/0000-0003-4769-4479

References

Aarsand, P., and H. Melander. 2016. "Appropriation through Guided Participation: Media Literacy in Children's Everyday Lives." *Discourse, Context and Media* 12: 20–31.

Albinsson, A. 2016. "De va svinhögt typ 250 kilo": Förskolebarns mätande av längd, volym och tid i legoleken. ["It was super high like 250 kg": Preschool children's measuring of length, volume and time in Lego play]. Diss lic., University of Linköping.

Bäckman, K. 2015. "Matematiskt gestaltande i förskolan." [Mathematical Formation in Preschool]. PhD diss., University of Åbo Akademi.

Baroody, A. J., M. L. Lai, and K. S. Mix. 2006. "The Development of Young Children's Early Number and Operation Sense and Implications for Early Childhood Education." In *Handbook of Research on the Education of Young Children 2*, edited by B. Spodek, and O. N. Saracho, 187–221. London: Routledge.

Caldera, Y. M., A. M. Culp, M. O'Brien, R. T. Truglio, M. Alvarez, and A. C. Huston. 1999. "Children's Play Preferences, Construction Play with Blocks, and Visual-Spatial Skills: Are they Related?" *International Journal of Behavioral Development* 23 (4): 855–872. doi:10.1080/016502599383577.

Casey, B. M., N. Andrews, H. Schindler, J. E. Kersh, A. Samper, and J. Copley. 2008. "The Development of Spatial Skills through Interventions Involving Block Building Activities." *Cognition and Instruction* 26 (3): 269–309. doi:10.1080/07370000802177177.

Clements, D. H. 2001. "Mathematics in the Preschool." *Teaching Children Mathematics* 7 (5): 270–275.

Doverborg, E., and I. Pramling Samuelsson. 2000. "To Develop Young Children's Conception of Numbers." *Early Child Development and Care* 162 (1): 81–107. doi:10.1080/0300443001620107.

Edens, K. M., and E. F. Potter. 2013. "An Exploratory Look at the Relationships among Math Skills, Motivational Factors and Activity Choice." *Early Childhood Education Journal* 41 (3): 235–243. doi:10.1007/s10643-012-0540-y.

Elia, I., A. Gagatsis, and M. van den Heuvel-Panhuizen. 2014. "The Role of Gestures in Making Connections between Space and Shape and their Verbal Representations in the Early Years: Findings from a Case Study." *Mathematical Education Research Journal* 26: 735–761. doi:10.1007/s13394-013-0104-5.

Ferrara, K., K. Hirsh-Pasek, N. S. Newcombe, R. M. Golinkoff, and W. Shallcross Lam. 2011. "Block Talk: Spatial Language During Block Play." *Mind Brain and Education* 5 (3): 143–151. doi:10.1111/j.1751-228X.2011.01122.x.

Gejard, G. 2014. "Jag kan göra matte å minus å plus": Förskolebarns och pedagogers deltagande i matematiska aktiviteter. ["I can do math and minus and plus": Pre-school children and teachers' participation in mathematical activities]. Diss. lic., Uppsala universitet: Studia Didactica Upsaliensia 7.

Goffman, E. 1978. "Response Cries." *Language* 54 (4): 787–815. doi:10.2307/413235.

Goodwin, C. 2000. "Action and Embodiment within Situated Human Interaction." *Journal of Pragmatics* 32 (10): 1489–1522. doi:10.1016/S0378-2166(99)0096-X.

Goodwin, C. 2007. "Participation, Stance and Affect in the Organization of Activities." *Discourse & Society* 18 (1): 53–73.

Goodwin, C. 2013. "The Co-Operative, Transformative Organization of Human Action and Knowledge." *Journal of Pragmatics* 46 (1): 8–23.

Goodwin, C., and M. H. Goodwin. 2004. "Participation." In *A Companion to Linguistic Anthropology*, edited by A. Duranti, 222–244. Malden, MA: Blackwell Publishing.

Heinemann, T., A. Lindström, and J. Steensig. 2011. "Addressing Epistemic Incongruence in Question-Answer Sequences through the Use of Epistemic Adverbs." In *The Morality of Knowledge in Conversation*, edited by T. Stivers, L. Mondada, and J. Steensig, 107–130. Cambridge: Cambridge University Press.

Jefferson, G. 2004. "Glossary of Transcript Symbols With an Introduction." In *Conversation Analysis: Studies from the First Generation*, edited by G. Lerner, 13–31. Amsterdam: John Benjamins Publishing Company.

Jirout, J. J., and N. S. Newcombe. 2015. "Building Blocks for Developing Spatial Skills: Evidence from a Large, Representative U.S. Sample." *Psychological Science* 26 (3): 302–310. doi:10.1177/0956797614563338.

Martin, C., and A.-C. Evaldsson. 2012. "Affordances for Participation: Children's Appropriation of Rules in a Reggio Emilia School." *Mind, Culture, and Activity* 19 (1): 51–74. doi:1080/10749039.2011.632049.

Mondada, L. 2009. "Video Recording Practices and the Reflexive Constitution of the Interactional Order: Some Systematic Uses of the Split-Screen Technique." *Human Studies* 32: 67–99. doi:10.1007/s10746-009-9110-8.

NCTM (National Council of Teachers of Mathematics). 2000. *Principles and Standards for School Mathematics*. Reston, VA: NCTM, 2000.

Ness, D., and S. J. Farenga. 2007. *Knowledge Under Construction: The Importance of Play in Developing Children's Spatial and Geometric Thinking*. New York: Rowman & Littlefield.

Ramani, G. B., E. Zippert, S. Schweitzer, and S. Pan. 2014. "Preschool Children's Joint Block Building During a Guided Play Activity." *Journal of Applied Developmental Psychology* 35 (4): 326–336. doi:10.1016/j.appdev.2014.05.005.

Reis, M. 2011. "Att ordna, från ordning till ordning. Yngre förskolebarns matematiserande." [To Order, from Order to Order. Toddler's Mathematizing]. PhD diss., University of Gothoburgensis.

Sarama, J., and H. D. Clements. 2009. "Building Blocks and Cognitive Building Blocks: Playing to Know the World Mathematically." *American Journal of Play* 1 (3): 313–337.

Seo, K.-H. 2003. "What Children's Play Tells us about Teaching Mathematics." *Young Children* 58 (1): 28–34.

Sfard, A. 2008. *Thinking as Communicating: Human Development, the Growth of Discourses, and Mathematizing.* Cambridge: Cambridge University Press.

Sfard, A. 2009. "What's all the Fuss About Gestures? A Commentary." *Educational Studies in Mathematics* 70 (2): 191–200.

Skoumpourdi, C. 2016. "Different Modes of Communicating Geometric Shapes, through a Game, in Kindergarten." *International Journal for Mathematics Teaching and Learning* 17 (2): 1–23.

Trawick-Smith, J., S. Swaminathan, B. Baton, C. Danieluk, S. March, and M. Szarwacki. 2016. "Block Play and Mathematics Learning in Preschool: The Effects of Building Complexity, Peer and Teacher Interactions in the Block Area, and Replica Play Materials." *Journal of Early Childhood Research* 1 (16): 1–16. doi:10.1177/1476718X16664557.

van Oers, B. 2014. "The Roots of Mathematizing in Young Children's Play." In *Early Mathematics Learning. Selected Papers of the POEM 2012 Conference,* edited by C. Benz, B. Brandt, U. Kortenkamp, G. Krummheur, S. Ladel, and R. Vogel, 111–123. New York: Springer.

Appendix

Transcription conventions

(Jefferson, 2004, for embodied actions Mondada, 2009)

[ ]	Overlapping talk
=	Equal signs indicate no break or gap between the lines.
(0.8) (.)	Numbers in parentheses indicate silence. A dot in parentheses indicates a micropause.
. , ¿ ?	The punctuation marks indicate intonation. The period indicates falling intonation, the comma continuing intonation, the inverted question mark slightly raising intonation, and the question mark indicates rising intonation.
::	Colons are used to indicate prolongation or stretching of the immediately prior sound.
-	A hyphen after a word indicates a cut-off or self-interruption.
word	Underlining indicates some form of stress or emphasis. The more the underlining the greater the emphasis.
WOrd	Especially loud talk is indicated by upper case.
° °	The degree signs indicate that the talk between them was quieter than its surrounding talk.
↑	The up arrow marks a sharp rise in pitch.
< >	Left/right carats indicate that the talk between them is slowed down.
.h .hh	Hearable inbreaths are shown with a ".h" – the more h's the more inbreath.
( )	Empty parentheses indicate that something is being said but no hearing can be achieved.
* *	Gestures and actions descriptions are delimited between two identical symbols (one symbol per participant) and are synchronized with corresponding stretches of talk.
–>	Gesture or action described continues after excerpt's end.
–> –>	Gesture or action described continues across subsequent lines until the same symbol is reached.
LISA	Name in upper-case indicates pedagogue
Hanna	Name in lower-case indicates child
hanna	Participant doing the gesture is identified when s/he is not the speaker.

4   Affective-motivational aspects of early childhood teacher students' knowledge about mathematics

Oliver Thiel and Lars Jenssen

ABSTRACT

One goal among many in early childhood education and care (ECEC) is that children develop positive attitudes towards mathematics. In order to support this development, teachers need professional competence in both cognitive and affective-motivational facets. Teacher students need to develop this professional competence during their initial teacher education. This paper investigates how teacher students' age, *mathematics self-efficacy* (MSE), and *mathematics anxiety* (MA) affect their knowledge about mathematics and mathematics education in ECEC settings. A quantitative cross-sectional study with a paper-and-pencil questionnaire is used with a convenience sample from Norway. While MA affects students' knowledge as expected from previous studies, our findings show that MSE has a net suppressor effect on MA, revealing a reversal paradox. Higher MSE leads to lower MA which leads to better achievement in the exam that was used to measure students' knowledge. However, when MA is controlled, higher MSE causes lower achievement. Furthermore, we found that MA is affected by teacher students' age. Older teacher students and part-time students with experiences from working in ECEC institutions are less anxious about mathematics. This finding suggests that positive experiences with mathematics in daily life and pre-school situations can help students overcome their MA.

Introduction

Most people agree that mathematics is important not only in science, business, and engineering but also in everyday life. Both doing mathematics and thinking mathematically are important, but different (Devlin 2012). Doing mathematics includes many everyday activities such as doing arithmetic while shopping or measuring when preparing a meal. Mathematical thinking is a specific way of reasoning, analysing structures, discovering patterns, and making connections (Devlin 2011, 59). Even young children start to think mathematically when learning to understand their world (Nakken and Thiel 2014, 20)

Many people have negative feelings towards mathematics (Neale 1969; Larkin and Jorgensen 2016). Attitudes and feelings develop from experiences. They are learned through dynamic interaction with the environment, where attitudes guide approaches to and avoidance of the subject (Metje, Frank, and Croft 2007; Ernest 2011). Negative beliefs often persist because of a lack of experiences that contradict them (Eiser et al. 2003). Early positive experiences of efficacy will reinforce themselves and lead to self-efficacy and positive attitudes, while early negative experiences of failure can lead to fear that is difficult to dispel later (Bandura 1977). Therefore, many curricula for early childhood education and care (ECEC) aim for children to have positive experiences with mathematics. For example, the official English version of the Norwegian framework plan for the content and tasks of kindergartens says: 'By engaging with quantities, spaces and shapes, kindergartens shall enable the children to … . find pleasure in mathematics' and staff shall 'encourage the children to be curious, find pleasure in mathematics and take an interest in mathematical relationships' (Norwegian Directorate for Education and Training 2017, 53–54).

Teachers need a positive attitude towards mathematics to achieve these goals. A teacher 'cannot share his enthusiasm when he has no enthusiasm to share' (Polya 1977, 1). Teachers' attitudes have an impact on children's mathematical learning (Karp 1991; Beilock et al. 2010). Teachers who have negative emotions towards mathematics may choose less effective mathematical activities or present them in a less effective manner. They may also be less supportive of children who engage in mathematical thinking (Gunderson et al. 2013). Bates, Latham, and Kim (2013) found that many ECEC teacher students in the Midwest of the United States have negative feelings towards mathematics. This matches our subjective experience from Norway, but there is little quantitative research regarding prospective ECEC teachers' attitudes towards mathematics in Norway and other Nordic countries (Alvestad et al. 2009).

How do teacher students' feelings affect their professional development and how can those feelings be changed. As mathematics teacher educators, we are interested in the factors that affect our teacher students' development of professional competence. Therefore, we wanted to investigate how affective-motivational aspects of teachers' professional competence affect students' acquirement of knowledge about mathematics. We have chosen to target mathematics self-efficacy (MSE) and mathematics anxiety (MA), which are the two most important aspects with a known impact on students' mathematics achievement (cf. Hembree 1990; Ho et al. 2000; Lee and Stankov 2013; Parker et al. 2014; Jenßen et al. 2015b; Oppermann, Anders, and Hachfeld 2016). As well, we are interested in how MSE and MA interact as they impact on ECEC teacher students' learning. How ECEC teacher students' age might affect beliefs, attitudes, and emotions are also investigated. The supposition is that the older a teacher student is, the more experiences they have.

Theoretical framework

During their studies, teacher students develop their professional competence in mathematics and mathematics education. Besides performance, knowledge, and situation-specific skills, beliefs, attitudes, and emotions are important facets of professional competence (Blömeke, Gustafsson, and Shavelson 2015). There is not yet an agreed definition of the term 'attitudes' (Hofer and Pintrich 2002). Fishbein and Ajzen (1975, 6) define attitude as 'a learned predisposition to respond in a consistently favourable or unfavourable manner with respect to a given object', while Fazio (1995) sees attitude as an association between an object and the evaluation of this object. We avoid this unprecise concept when focussing on MSE and

MA. Those two affective-motivational concepts have precise definitions and are the two most important facets affecting all students' learning and performance in mathematics.

Mathematics self-efficacy (MSE)

Self-efficacy is a belief that is related to the ECEC teacher student's self. It can be defined as a 'belief in one's capabilities to organise and execute the courses of action required to produce given attainments' (Bandura 1977, 193). As well as a person's general self-efficacy, there are domain-specific self-efficacies (Bandura 1986). While general self-efficacy is not strongly related to performance in mathematics (Benson 1989; Cooper and Robinson 1991), this is not the case for MSE. MSE is defined as 'a situational or problem-specific assessment of an individual's confidence in her or his ability to successfully perform a particular math-related task or problem' (Hackett and Betz 1989, 262). Among the fifteen non-cognitive variables that were measured in the PISA 2003 study, MSE had the strongest impact on students' mathematics achievement (Lee and Stankov 2013). MSE is a predictor of later achievement even if prior achievement is controlled (Parker et al. 2014). However, both studies addressed 15-year-old students. International studies (Bates, Latham, and Kim 2011; Tsamir et al. 2015) found moderate correlations between ECEC pre-service teachers' MSE and their mathematical performance. Oppermann, Anders, and Hachfeld (2016) showed that MSE mediates between ECEC teachers' mathematical content knowledge and their ability to recognise mathematical content in play situations.

Mathematics anxiety (MA)

Negative feelings towards mathematics in their strongest form are 'feelings of tension and anxiety that interfere with the manipulation of mathematical problems in a wide variety of ordinary life and academic situations' (Richardson and Suinn 1972, 551). This is MA. MA can be examined on a cognitive (e.g. thoughts about own failure), affective (e.g. fear), physiological (e.g. increased muscle tension), or on a behavioural (e.g. avoidance) level (cf. Bessant 1995). The negative impact that MA has on mathematics achievement is well known (Hembree 1990; Ho et al. 2000). The meta-analysis by Ma (1999) has shown a moderate negative correlation between MA and performance in mathematics among primary and secondary students. This has been shown for ECEC teachers in Germany, too (Jenßen et al. 2015b). Even though ECEC teacher education in Germany and Norway are very different, we expect that MA will negatively affect Norwegian ECEC teacher students' acquirement of mathematical knowledge.

There are two possible causal directions between MA and poor mathematics performance. Most likely, it is a bidirectional relationship, in which MA and mathematics achievement can influence one another cyclically (Carey et al. 2016). In our case, we are following the Debilitating Anxiety Model suggesting that MA can impact performance because persons with increased MA may avoid mathematics-related situations (Hembree 1990; Chipman, Krantz, and Silver 1992).

ECEC teachers' mathematical knowledge for teaching

According to the Norwegian framework plan for kindergartens, staff in Norwegian ECEC institutions shall, among other things, 'inspire the children's mathematical thinking, ...

create opportunities for mathematical experiences by enriching the children's play and day-to-day lives with mathematical ideas and in-depth conversations, stimulate and support the children's capacity for and perseverance in problem-solving' (Norwegian Directorate for Education and Training 2017, 54). In order to do this, ECEC professionals need to have adequate knowledge for teaching mathematics in ECEC settings. Hill, Rowan, and Ball (2005, 373) define mathematical knowledge for teaching (MKT) as 'the mathematical knowledge used to carry out the *work of teaching mathematics*' (emphasis in the original). Based on many studies, Ball, Thames, and Phelps (2008) developed a framework of MKT that identifies six different domains – three related to subject matter knowledge and three related to pedagogical content knowledge.

Domains related to subject matter knowledge are common content knowledge (CCK), horizon content knowledge (HCK), and specialised content knowledge (SCK). CCK is 'the mathematical knowledge and skill used in settings other than teaching' (Ball, Thames, and Phelps 2008, 399). The Norwegian Mathematical Council measures first-year students' mathematical knowledge in programmes that include mathematics. Primary teacher students always score significantly lower than secondary teacher students (Nortvedt and Bulien 2016, 16). ECEC teacher students do not participate in the Mathematical Council exercise, but a study by the Norwegian Agency for Quality Assurance in Education (NOKUT) shows that it is reasonable to assume that their mathematical CCK is the same or lower than that of primary teacher students (Lid, Pedersen, and Damen 2018, 28 and 59).

HCK is 'an awareness of how mathematical topics are related over the span of mathematics included in the curriculum' (Ball, Thames, and Phelps 2008, 403). For early childhood mathematics teachers, it is important to have knowledge about mathematics in grade one and two of primary school. Results from the Programme for the International Assessment of Adult Competencies (PIAAC) in Norway show that less than 5% of people in education have a proficiency in numeracy below level 1; among people with higher education, it is less than 1% (Statistics Norway 2013). Lower primary school mathematical knowledge is on level 1 and useful in many everyday life situations. Therefore, it can be assumed that student teachers in this study most likely have adequate HCK.

SCK is 'the mathematical knowledge and skill unique to teaching' (Ball, Thames, and Phelps 2008, 400). This is the domain in which we are most interested.

> Teachers … . must be able to talk explicitly about how mathematical language is used (e.g. how the mathematical meaning of *edge* is different from the everyday reference to the edge of a table); how to choose, make, and use mathematical representations effectively … .; and how to explain and justify one's mathematical ideas (Ball, Thames, and Phelps 2008, 400, emphasis in the original).

Dunekacke, Jenßen, and Blömeke (2015) and Oppermann, Anders, and Hachfeld (2016) showed that mathematical content knowledge is generally beneficial for ECEC teachers' sensitivity to mathematical content in ECEC settings. There is still no general consensus regarding what SCK is required by ECEC teachers.

Domains related to pedagogical content knowledge are knowledge of content and students (KCS), knowledge of content and teaching (KCT), and knowledge of content and curriculum (KCC). KCS is 'knowledge that combines knowing about students and knowing about mathematics. Teachers must anticipate what students are likely to think and what they will find confusing'. It 'requires an interaction between specific mathematical understanding and familiarity with students and their mathematical thinking' (Ball, Thames, and Phelps 2008, 401). To this domain belongs knowledge about children's understanding and development of mathematical concepts (e.g. Piaget and Szeminska 1941; Gelman and Gallistel 1978; Vygotsky 1978; Fuson and Hall 1983; van Hiele 1986; Dehaene 2011).

KCT

> combines knowing about teaching and knowing about mathematics. Many of the mathematical tasks of teaching require a mathematical knowledge of the design of instruction, …. . an interaction between specific mathematical understanding and an understanding of pedagogical issues that affect student learning (Ball, Thames, and Phelps 2008, 401).

KCT needed by ECEC teachers is very different from primary and secondary school teachers' KCT because the 'work of teaching mathematics' in a Norwegian kindergarten is different from school, as Mosvold et al. (2011, 1807) describe: 'In a typical Norwegian kindergarten, there are no mathematics lessons, no mathematics textbooks, no board where the teacher presents anything, and no traditional classroom with desks. The children's learning takes place in everyday activities and play situations'. While teachers in primary and secondary schools usually start with the mathematical content and then use their pedagogical KCT to plan a lesson, pre-school and kindergarten have a more holistic approach. Tirosh et al. (2011, 116) observed that ECEC teachers 'engaged with challenging tasks, addressing first subject matter knowledge, and then related pedagogical content knowledge issues. This led to a gradual interweaving of knowledge and practice'. They use this interweaved knowledge of mathematics content, pedagogy, and children's development to create rich learning environments and to exploit teachable moments in everyday life and play situations.

KCC is curricular knowledge. The section about mathematics (called 'Quantities, spaces and shapes') in the Norwegian ECEC curriculum (Norwegian Directorate for Education and Training 2017, 53–54) has fewer than 300 words. Thus, it is not hard to remember the words but difficult to interpret and apply what is meant. Østrem et al. (2009) investigated the implementation of the curriculum from 2006. They revealed that many ECEC teachers have insufficient knowledge about content and importance of some of the mathematical domains that are mentioned in the curriculum. This is still the same today (Bjørseth 2017).

The model developed by Ball, Thames, and Phelps (2008) shows that MKT is complex and multidimensional. Whether the presented domains 'are the right ones is not most important. Likely they are not' (Ball, Thames, and Phelps 2008, 403). There are different approaches to describe mathematical knowledge in teaching (cf. Rowland and Ruthven 2011). To discuss all approaches in detail would go beyond the scope of this article. Despite the research that was completed in the last decades, Deborah L. Ball still admits that 'the teaching profession lacks a shared codified professional knowledge' (Arbaugh et al. 2015, 436).

There is growing interest in applying Ball et al.'s framework in the Norwegian context (Fauskanger et al. 2012; Mosvold and Fauskanger 2013; Jankvist et al. 2015). Mosvold et al. (2009) translated the instrument developed by Hill et al. (2007) to measure MKT in Norway. The results of their study question the instrument's validity for the Norwegian context (Mosvold and Fauskanger 2015). More important, it is not applicable to ECEC teachers (Mosvold et al. 2011). Thus, there is still 'little evidence regarding what kind of mathematical knowledge for teaching that is needed by kindergarten teachers' (Mosvold et al. 2011, 1802). In Norway, there is a curriculum for ECEC teacher education (Norwegian Directorate for Education and Training 2012). It states that the students shall acquire knowledge about mathematical domains that are relevant for children (p. 19), but it does not mention any specific mathematical content. Nevertheless, everyone who teaches mathematics for prospective ECEC teachers has a notion of what knowledge is needed, every mathematics textbook for ECEC teacher students provides a curriculum that is built on the available evidence and theoretical considerations, and every mathematics exam for ECEC teacher students is designed to measure to what degree the students have gained the knowledge that was taught in the teacher training course.

The instrument by Hill et al. (2007) uses – like most standardised test instruments – multiple-choice items, but Schoenfeld (2007), as well as Buchholtz et al. (2013), argues that open-ended essay questions are more appropriate to measure students' deeper understanding of the subject matter. Cahill et al. (2018) investigated the relationship between student attitudes and student learning. They used both concept inventories and exam averages as measures of conceptual knowledge and found that the overall pattern of results was the same for both measures. Slepkov and Shiell (2014) directly compared constructed-response (CR) questions with multiple-choice testing. CR problems require the students to demonstrate their integration of a wide range of skills and concepts. The answer is interpreted by an expert who gauges its level of correctness. CR questions are comparable to the type of questions used in the written mathematics exams in ECEC teacher education. Slepkov and Shiell (2014) dispelled notions that multiple-choice tests and CR questions about the same content measure distinctly different constructs.

In his meta-analysis about the correlation between MA and mathematics achievement, Ma (1999) compared 26 individual studies. Of those studies, 15 used standardised tests to measure students' mathematics achievement, 1 used a test developed by the researchers, and 10 used the mathematics teacher grading. For a small-scale study like ours, it seems reasonable to use the teacher grading of the teacher students' written exam that contains three open-ended essay questions as a sufficient measure of students' MKT instead of developing a standardised multiple-choice test.

Present study

The research questions

The aim of the study presented here is to answer the following research questions:

1. How do prospective ECEC teacher's age, MSE, and MA interact with each other in affecting students' knowledge about mathematics and mathematics education in ECEC settings?
2. Which of MA and MSE affect ECEC teacher students' knowledge about mathematics and mathematics education in ECEC settings more strongly?
3. Does MSE affect students' knowledge mostly indirectly via MA, or is there a direct effect as well?
4. How does students' age affect their MA?

The model to be tested

We tested a structural equation model (SEM), i.e. a model that describes the directed dependencies among a set of variables. With SEM, it is possible to impute relationships between unobserved constructs (latent variables) from observable variables (Hancock 2003). In our model, we use MSE and MA as two latent variables and the students' age and exam grade as observable variables. Since the aim of our study is to analyse the impact that ECEC teacher students' MSE and MA have on their performance in the exam, we assume that both constructs directly affect students' exam grade. Why this is reasonable has been discussed in the previous sections.

Jain and Dowson (2009) found that self-regulation and self-efficacy have a direct negative impact on MA among secondary school students. In a study with elementary pre-service teachers, Swars, Daane, and Giesen (2006) showed that higher mathematics teacher efficacy leads to lower MA. McMullan, Jones, and Lea (2012) found among undergraduate nursing students that MSE correlates strongly and negatively with MA and that students' numerical ability correlates moderately with MSE and MA. None of these studies has been made with prospective ECEC teachers, but since the same dependency was found in different populations, we expect to find the same in our sample. Thus, we assume that ECEC teacher students' MSE directly affects their MA. That means, we assume that MSE has not only a direct impact on students' performance but also an indirect impact that goes via MA (i.e. lower MSE leads to higher MA which causes lower performance). With an SEM, it is possible to investigate whether the direct or indirect impact is stronger. To our knowledge, this has not been done before.

Age is a factor that impacts MA (Dowker, Sarkar, and Looi 2016). Since we have students of different age in our sample, we have to consider this. Research with children and adolescents has shown that older students experience more MA than younger ones (Meeks 1997; Jain and Dowson 2009; Hill et al. 2016). Jameson and Fusco (2014) have compared adult learners with traditional undergraduate students. They found the same relationship between age and levels of MSE and MA: older students had lower levels of MSE and higher levels of MA than their younger peers. However, university mathematics students are not comparable to teacher students, and Thiel (2010) found a different pattern among ECEC teachers. Older ECEC teachers feel more often open and younger ones more often reluctant towards mathematics. We assume that students' age directly affects their MA. This means that the students' age has an indirect, rather than a direct, impact on the exam. Figure 1 shows a graphical representation of the model.

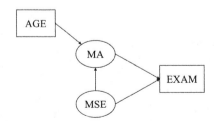

Figure 1. The model to be tested. AGE: students' age; EXAM: students' grade in the written exam about mathematics and mathematics education in ECEC. Rectangular boxes represent observable variables and oval boxes represent latent variables. Arrows represent theoretical causal impacts.

Method

Participants and procedure

The study reported here is a small-scale quantitative cross-sectional study with a paper-and-pencil questionnaire and a convenience sample. This means we do not have a random sample that is representative for the whole country but one that we had the opportunity to obtain. In the school year 2014/15, all ECEC teacher students who attended a course in ECEC mathematics education at one Norwegian university college were asked to participate in the study. This university college educates about 14% of all Norwegian ECEC teachers (Norwegian Centre for Research Data 2018). The response rate was 64% – this means 225 persons (64% of the population of students asked) filled in the questionnaire in April or May 2015, one week before the written exams in ECEC mathematics education. The assessment took place during regular instruction times. Four students had to be eliminated from the study due to blank questionnaires. The sample represents about 9% of all ECEC teacher students who completed a course in ECEC mathematics education in Norway in the school year 2014/15. All participants provided informed consent, and all responses were anonymous to align with institutional review board requirements. The data were digitalised and processed in Berlin, Germany, by researchers who had never interacted with the participants in person.

The sample contains six classes with full-time students: one class with emphasis on music, drama, arts, and crafts; one class with emphasis on nature and outdoor life; and four classes with no special emphasis. In addition, there were three classes with part-time students. Independent of their status as a full- or part-time student, all participants had undergone the equivalent of two years of full-time study in teacher education, including one year with about 8 ECTS[1] credits in ECEC mathematics education (about 30 hours in class and 170 hours self-study).

Measures

MSE: MSE was surveyed with four items developed by Jerusalem and Satow (1999) based on well-established scales assessing teachers' general self-efficacy (Schwarzer and Jerusalem 1995) that were adapted for the target population of the present study. The items are listed in Table A2 in the Appendix. The items have been translated to Norwegian observing scientific conventions for test translation (International Test Commission 2005), resulting in a two-step process. The items were first translated to Norwegian by a native German speaker who is fluent in Norwegian language and culture since he has lived in Norway for many years. Then, the items were translated back to German by a native Norwegian speaker who has lived in Germany for many years. Thus, we could assure that the meaning of the items has not been altered under translation. As with the original scale, participants had to rate the items on a 4-point Likert scale ranging from 'strongly disagree' to 'strongly agree'. A high score indicates high MSE. Reliability of the scale can be seen as sufficient (Cronbach's α = .86). In the SEM, MSE is modelled as a latent variable using the four items as indicators.

MA: The Mathematics Anxiety Scale-Revised (MAS-R) by Bai et al. (2009) was used to measure students' MA. The questionnaire contains 14 items. Six items are statements about positive thoughts (e.g. 'I find math interesting'). They represent the cognitive component of MA. The remaining eight items are statements about negative emotions (e.g. 'Mathematics makes me feel nervous'). They represent the affective component of MA. A list of all items is given in Table A3 in the Appendix. As with the original scale, participants had to rate on a 5-point Likert scale. The positive statements have been scored in reverse so that a high score indicates high anxiety. The questionnaire was translated observing scientific conventions for test translation (International Test Commission 2005) using the same procedure of translation as for the MSE items. The Norwegian version has a high reliability (Cronbach's $\alpha = .93$). An exploratory factor analysis shows that the positive and negative statements represent the two different components of MA (see Table A3 in the Appendix). Therefore, we used parcels as indicators of the latent variable MA (cf. Coffman and MacCallum 2005), one parcel MA+ for the cognitive component and one parcel MA− for the affective component of MA.

EXAM: As a sufficient measure for the prospective ECEC teachers' MKT, we used the student's grade in the final written exam in mathematics and mathematics education in ECEC settings. Grades are A (excellent), B (very good), C (good), D (rather good), E (sufficient), and F (inadequate). We coded A as 6 and F as 1 so that higher values represent better performance. In the SEM, the exam grade is used as an observed dependent variable. The exam was a 3-hour written assessment of specialised mathematical content knowledge and mathematics pedagogical content knowledge, both focused on ECEC settings. The tasks were developed in collaboration with all mathematics teachers at the involved university college. They agreed that the exam content covers a representative sample of the content of students' required reading which was Devold (2008, 2010), Nakken and Thiel (2014), Reikerås and Fauskanger (2008), and Valle (2008). Table A4 in the Appendix shows an English translation of the exam tasks.

Task 1 focused on conceptual mathematical content knowledge. In part (a), the candidate had to explain four different concepts of SCK. In part (b), she or he had to apply KCT in order to relate the concepts to ECEC settings. A maximum of 7.5 points was given for each of the four concepts (variables T11, T12, T13, and T14). Task 2 focused on mathematics pedagogical content knowledge, especially (a) KCT about creating a mathematical learning environment and (b) KCS about fostering children's learning in this learning environment. A maximum of 15 points was given for T2a and a maximum of 30 points for T2b. In order to solve task (3a), the candidate had to identify an abstract mathematical concept in a given realistic situation from kindergarten. A maximum of 5 points was given for T3a. In task (3b), the candidate had to explain the concept mathematically. Task (3c) was about applying the concept to children's life. A maximum of 10 points was given to T3b and T3c each.

The three tasks 1, 2, and 3 together cover a wide range of MKT, namely explaining terms and concepts, arranging learning environments and developing materials for early childhood mathematics, and interpreting children's statements. The exam was gauged by three mathematics teachers collaboratively. The teachers were not involved in the later statistical analyses. Each teacher evaluated one task. Afterwards, they discussed their judgements and came to an agreement about the final score. The maximum score was 100 points. Points were converted to grades in the following way: 90 points or more is grade A, 80 or more but less than 90 points is grade B, 60 or more but less than 80 points is C, 50 or more but less than 60 points is D, 40 and more but less than 50 points is E, and less than 40 points is F. In contrast to standardised tests, the exam contains only nine items, which have quite different weights. This is typical for this kind of exam. Taking this into account, the internal consistency estimate of the reliability of the exam scores is rather good (Cronbach's α = .72). Table A5 in the Appendix shows variances, correlations, and covariances between the items. Inter-item correlations are rather low. This shows that MKT is a multidimensional construct.

Statistical analyses

The fit of the model to the data was examined through a Confirmatory Factor Analysis (CFA). CFA is a statistical procedure to test how well the measured variables (i.e. the items from the questionnaire) represent the constructs (i.e. MSE and MA). The Root Mean Square Error of Approximation (RMSEA) and the Standardised Root Mean Residual (SRMR) are fit indices that estimate how well the model reproduces the observed covariance matrix. Estimates less than .05 point to a very good fit (Hu and Bentler 1999). The Comparative Fit Index (CFI) assesses to what extent a model reproduces the observed covariance matrix better than a baseline model that is assuming all observed variables are uncorrelated. Estimates larger than .95 point to an acceptable fit. Analyses were done in *Mplus* 7.1 (Muthén and Muthén 2014) using a robust maximum likelihood estimator and taking into account that students were clustered in classes with different teachers (TYPE = COMPLEX procedure in *Mplus*). When using a robust maximum-likelihood estimator, standard errors and a chi-square test statistic are robust to non-normality and non-independence of the observations. To handle missing data, full-information-maximum-likelihood procedures were used. This is a common estimation method in CFA models. It is based on all information contained in the response pattern. Usually, the observed variables have to be continuous and normally distributed, but if the responses are ordered categorical variables – like in our case – one can treat them as continuous variables if they are arranged numerically in ascending order. The model parameters can be estimated without examining the categorical nature of the variables (Suh 2015).

Results

Raw scores

The students' ages are between 20 and 54 years with a mean of 27.2. The sample includes both full-time (N = 138) and part-time (N = 83) students. Part-time students are significantly older (33.5 vs. 23.5 years, $p < .001$, see Table 1). They match the definition of 'non-traditional student' by the U.S. National Centre for Education Statistics (NCES) (Choy 2002; Jameson and Fusco 2014). However, some of the full-time students match that definition, too. We define a 'non-traditional student' as one who has any of the following characteristics: is a part-time student, has no high-school diploma, or is at least 27 years old. The sample contains 120 (53%) traditional and 105 (47%) non-traditional students.

Grades in the written exam happened to be normally distributed. The average grade in the sample is not significantly different from the average grade of all students at the university college. In the sample, non-traditional students have significantly better grades than traditional students ($p < .01$). The same is true for part-time versus full-time students.

Table 1. Raw scores of observed and aggregated variables

Variable	Sample	M	SD	Min	Max	ΔM	t	df	p
							t-test		
AGE		27.24	7.07	20	54				
	Full-time	23.50	3.26	20	39	−9.95	−11.663	101.8[a]	.000
	Part-time	33.45	7.35	22	54				
MSE		7.68	2.62	4	14				
	Full-time	7.52	2.51	4	14	−.39	−1.029	195	.305
	Part-time	7.91	2.77	4	14				
MA		39.24	11.41	16	67				
	Full-time	41.46	10.80	18	67	5.42	3.354	194	.001
	Part-time	36.04	11.57	16	66				
EXAM 2015		3.81	1.22	1	6				
	Full-time	3.59	1.07	1	6	−.65	−2.887	122	.005
	Part-time	4.24	1.39	1	6				
EXAM 2016		3.67	1.26	1	6				
	Full-time	3.50	1.17	1	6	−.55	−2.682	167	.008
	Part-time	4.06	1.38	1	6				

[a]According to Levene's test, equal variances cannot be assumed for full- and part-time students' age. Therefore, Welch's *t*-test was chosen.

M: mean; *SD*: standard deviation; *Min*: minimum, *Max*: maximum; *ΔM*: difference of means for full-time and part-time students; *t*: t-score according to Student's *t*-test (Welch's *t*-test in the case of age since equal variances cannot be assumed); *df*: degrees of freedom; *p*: significance, i.e. likelihood that population means are equal; AGE: students' age in years; MSE: students' mathematical self-efficacy; higher scores represent higher self-efficacy; MA: students' mathematics anxiety; higher scores represent higher anxiety EXAM: students' grade (the best grade A is coded 6, the worst grade F is coded 1) in the written exam about mathematics and mathematics education in ECEC in the sample from 2015 and in the following year 2016.

The difference is about two-thirds of a grade or a half standard deviation. This could indicate that part-time students have been rated differently, but we think it represents a real difference in knowledge, since part- and full-time students have been assessed by the same examiners and the difference appears in the exam grades of the following year, too (see Table 1). Non-traditional full-time students' average grade is not significantly different from part-time students' average grade.

The MSE score can range from 4 to 16, but in this sample, the maximum score is 14. Item means are provided in Table A2 in the Appendix. The mean of the sum of all items $M_{MSE} = 7.68$ is significantly below the expected mean for a normal distribution ($p < .001$). There is neither a difference between traditional and non-traditional students nor between full-time and part-time students. Age and MSE are not correlated.

The MA can possibly range from 14 to 70. In our sample, the range is from 16 to 67. The most agreed negative item is 'I find math challenging'. Seventy-six per cent of all students (rather or totally) agree with this statement; only 7% disagree. The most agreed positive item is 'I think that I will use math in the future'. Even 84% of all students agree; only 5% disagree. The most disagreed positive item is 'Math is one of my favourite subjects'. Fifty-six per cent of the students disagree but 20% agree. The most disagreed negative item is 'Mathematics makes me feel uneasy'. Sixty-two per cent of the students disagree but 20% agree. Item means are reported in Table A3 in the Appendix. The mean of the sum of all items $M_{MA} = 39.24$ is significantly lower than average expected for a normal distribution ($p < .01$) and much lower than the mean that Jenßen et al. (2015b) found with the same instrument in a sample of German prospective ECEC teachers ($p < .001$). Non-traditional students have a significantly lower MA than traditional ones (36.56 vs. 41.77, $p < .01$). The same is true for part-time versus full-time students (see Table 1).

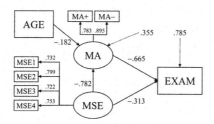

Figure 2. The SEM. EXAM: students' grade in the written exam about mathematics and mathematics education in ECEC settings; For MSE1, MSE2, MSE3, MSE4, MA+ and MA−, see Tables A1, A2, and A3 in the Appendix.

There is a weak negative correlation ($r_{01} = -.182$, $p < .05$) between age and MA. Students who are at least 27 years old have significantly lower MA than younger students (36.70 vs. 40.63, $p < .05$).

As expected from our model, we found a strong negative correlation between MSE and MA ($r_{12} = -.782$, $p < .001$), a moderate negative correlation between MA and the exam ($r_{1c} = -.420$, $p < .001$), and a weak positive correlation between MSE and the exam ($r_{2c} = .207$, $p < .05$). By looking closer, we discovered a negative net suppression because $0 < r_{2c} < r_{1c}r_{12}$ (cf. Cohen and Cohen 1975; Krus and Wilkinson 1986; Lewis and Escobar 1986). The SEM will be examined to explain this further.

The SEM

The model is shown in Figure 2. Variances, covariances, and correlations of all manifest variables are shown in Table A1 in the Appendix. The model fit is very good ($\chi^2(18) =$ 24.98, $p = .125$, RMSEA = .042 [.000–079], SRMR = .032, CFI = .996). The model explains 64.5% of the variance of MA. Since we assume a causal relation, this means that 61.1% of the variance of students' MA is caused by the variability of their MSE. Only 3.3% of MA's variance can be predicted by students' age. The model explains 21.5% of the variance of the exam grades. It is not surprising that this is only about one-fifth of the variance since we know that there are many additional factors affecting students' exam performance. MA explains 17.6% of the grades' variance. The absolute value .662 of the path coefficient β_{1c} from MA to EXAM is larger than the absolute value .420 of the correlation r_{1c} between MA and the exam grade because of the suppressor MSE. In total, MSE explains only 4.3% of the variance of the exam grades, but there is a much stronger indirect effect from MSE via MA to EXAM ($\beta_{21}\beta_{1c} = .521 > r_{2c}$). The path coefficient β_{2c} from MSE to EXAM reveals a reversal paradox (Messick and Van de Geer 1981): The direct effect has the opposite sign to the total effect. We will discuss this in the next section.

Discussion

MA has a stronger effect than MSE

The findings confirm our expectation that students' MA affects their exam performance negatively while their MSE has a positive effect. One of our research questions was: Which effect is stronger? Hackett and Betz (1989) found in their study with undergraduate psychology students that the effects that students' MA and MSE have on their perform-ance in the mathematics section of the American College Test have almost the same size. This is different in our sample. While the correlation between MA and EXAM is the same as found by Hackett and Betz (1989), the correlation between MSE and EXAM is much weaker. It does not surprise that students' belief in their mathematics skills correlates more with their actual mathematics skills than with their MKT. The reason for the weak correlation between MSE and the EXAM is the net suppression effect that we will discuss in the next section. More alarming is that the correlation between students' anxiety and the exam grades is that strong, even stronger than the average correlation that Ma (1999) found in his meta-analysis. Following the Debilitating Anxiety Model (Hembree 1990; Chipman, Krantz, and Silver 1992) that MA impacts per-formance because persons with increased MA avoid mathematics-related situations, this means that ECEC teacher students may avoid ECEC-related mathematics even though they have been taught that it is quite different from traditional school mathematics. This is a serious problem that has to be addressed in ECEC teacher education (cf. Bates, Latham, and Kim 2013).

MSE and the reversal paradox

A well-known example for net suppression is the apparent paradox that the number of fire-fighters involved in extinguishing a fire is positively correlated to the damage caused by the fire (Messick and Van de Geer 1981). The crucial third variable is the sever-ity of the fire. The worse the fire, the more fire-fighters will fight it. For fires of equal sever-ity, the correlation, in fact, has a reversed sign, i.e. more fire-fighters will reduce the damage. What does this mean in our case? Often it is argued that greater self-efficacy pro-vides more effort and persistence even if the students overestimate their capability (Pajares and Miller 1994). This is in accordance with the positive correlation between MSE and the exam grades that we have found in our sample. However, our findings show that anxiety plays an important role in this process. In our sample, the positive impact that MSE has on students' achievement goes mainly via MA: Students who have a stronger belief in their ability to accomplish mathematical tasks are less anxious about mathematics and hence more motivated to make an effort which enhances their performance in the exam. Stu-dents who do not believe in their ability are more anxious. This leads to avoidance and lower performance.

On the other hand, if MA is controlled, greater MSE leads to lower performance. Among students with an equal amount of MA, a stronger belief in one's mathematical abilities has a negative effect on the student's achievement in the exam. We can understand this in the following way: A student with greater MSE says: 'Although I have not solved mathematical problems for a long time, I am nevertheless able to solve such problems when I receive them' (see item MSE2 in Table A2). This belief might provoke poor prep-aration for the exam. This would be okay if the belief is in accordance with the student's actual abilities, but is a problem for students who overestimate their capabilities. More-over, the exam used in this study is not about mathematical problems but about ECEC mathematics pedagogical content knowledge. A lack of knowledge cannot be compensated by pure mathematics skills. On the other hand, students who do not feel anxious about mathematics but believe that their abilities are not good enough might put a lot of effort into the preparation for the exam in order to gain a better result. This sounds reasonable, but we do not know for sure if it is the case because we have not asked the students about their preparation for the exam.

In summary, our findings show that MSE has both positive and negative effects on students' achievement. Underestimating one's abilities leads to irrational anxiety, and overestimating one's abilities might cause poor preparation for the exam. Thus, it is important for ECEC teacher educators to help students develop a realistic MSE.

Age and the importance of experience

We found that older ECEC teacher students have slightly less MA than younger ones. The level of MA among the youngest ECEC teacher students is comparable to that of secondary school students. This reinforces earlier findings (Thiel 2010) but contradicts the findings by Jameson and Fusco (2014). Therefore, we have to explain why adult learners studying ECEC pedagogy are different from adult learners who study other subjects. Bekdemir (2010) has shown in a sample of pre-service primary school teachers that negative experiences with mathematics in primary and secondary school initially create and then raise MA over the years. During compulsory school, students' MA increases because the older school children had more opportunities 'to experience failure or the threat of it' (Dowker, Sarkar, and Looi 2016, 9). Younger teacher students enter university right after school and vividly recall their own unpleasant experiences with mathematics at school. They have the same level of MA as upper secondary students. But what happens to people who do not enter university right after school? We know in general that memories of unpleasant experiences fade as times passes by – even if the effect is lowered for people with increased anxiety (Walker, Yancu, and Skowronski 2014). Jameson and Fusco (2014, 314) found in normal university mathematics classes that adult learners' self-efficacy decreases and their MA increases as their age increases, but those students are indeed not comparable to the students in our sample. Many of our adult ECEC teacher students have jobs in ECEC institutions. In the university college, they are not confronted with formal mathematics but mathematical content knowledge for teaching in ECEC settings. We often observe that it is easier for them to relate the mathematical content to their experiences from ECEC institutions than it is for younger students to imagine what early childhood mathematics might be about. Therefore, we suppose that positive experiences with mathematics in daily life and ECEC situations may help students to understand early childhood mathematics better and, thus, to reduce their MA. The objective of an ongoing follow-up research project is to find out if this is the case.

Limitations

Since this is a small-scale study with a convenience sample, the representativeness of the present study is limited. Even though the sample represents 9% of the population, it does not reflect the full heterogeneity of prospective ECEC teachers in training. ECEC teacher education in Norway varies from university to university both in the amount of ECTS credits and content in ECEC mathematics education courses. Thus, the generalisability of our findings is limited to comparable groups represented in our sample. Future research should examine whether the same findings apply to other groups of prospective ECEC teachers (e.g. prospective ECEC teachers on the post-secondary level at vocational schools).

Another problem might be that we did not use a standardised test to measure students' MKT. Exam grades are not necessarily an objective representation of students' knowledge. Performance in an exam is affected by the candidate's current mood and affective-motivational factors, and the examiner's judgement can be affected by subjective factors like the candidate's parlance. Fortunately, the evaluation of written exams in Norway is anonymous. Thus, the grade is not affected by the examiner's subjective impression of the student.

Even though the validity of the content of the exam tasks is given and the exam has sufficient empirical reliability, not using a standardised test makes it difficult to compare our findings with other studies. We examined a German sample of prospective ECEC teachers using the same instruments for MSE and MA, but measuring students' knowledge with a standardised test for mathematical content knowledge (MCK) that was developed for the German research project KomMa (Jenßen et al. 2015a). We found the same suppressor effect in this as yet unpublished data (Jenßen, Thiel, Dunekacke, & Blömeke, forthcoming). Therefore, we believe that it is not a statistical artefact. However, more research is needed to investigate if this is a general characteristic of prospective ECEC teachers.

Due to numerical restrictions in the statistical methods, we could consider only a selected set of variables in our model. We have addressed more aspects of the same investigation in other articles (e.g. Blömeke, Thiel, and Jenßen 2017). Other characteristics, such as ECEC teachers' self-efficacy with respect to implementing mathematical activities, not only of mathematics as a subject, may be important, too. There may be ECEC teachers who are anxious about (formal) mathematics but positively value certain types of (playful) mathematical activities with children. We do not know how they would act. It is important to identify such conflicts so that it may be possible to address them in ECEC teacher education. This will be addressed in future studies.

Conclusion

Our research question was: How do prospective ECEC teacher's age, MSE, and MA interact with each other in affecting students' knowledge about mathematics and mathematics education in ECEC settings? The interactions are displayed in the SEM. It shows that it is not enough to focus on cognitive aspects only when looking at ECEC teachers' mathematical competence (Fröhlich-Gildhoff, Nentwig-Gesemann, and Pietsch 2011). Other studies showed that beliefs are important, too (Thiel 2010; Benz 2012; Dunekacke et al. 2016). Our findings indicate that affective-motivational aspects have to be considered as well – something that is often neglected in research on competence. Affective-motivational characteristics such as MA and self-efficacy affect students' knowledge assessed with a written exam. Hence, these topics should be addressed in ECEC teacher training.

We expected that MA affects students' exam performance negatively while MSE has a positive effect, and asked: Which effect is stronger? We found that MA has the stronger effect. MA is a serious problem that arises and accumulates during the years in school (Bekdemir 2010; Dowker, Sarkar, and Looi 2016). In ECEC, MA has two negative effects: During ECEC teacher education, it prevents students from acquiring the necessary mathematical knowledge that is taught. This is a clear result of our study. Later in pre-schools, ECEC teachers with MA tend to avoid mathematics in ECEC settings (Stipek 2017). How does students' age affect their MA? Yes, older students have slightly lower MA. Most of them are part-time students who already work in ECEC institutions and had the opportunity to make experiences with mathematics in ECEC settings. Therefore, we assume that positive experiences with mathematics in daily life and ECEC situations help students to overcome their MA. This might be a possible way to approach this problem in ECEC teacher training. Bates, Latham, and Kim (2011) recommend develop-ing a playful inquiry-based approach to students' own mathematics as it already exists in early childhood mathematics education. In this way, students could gain positive experi-ences which would be remembered both cognitively and emotionally. In our Norwegian University College, we have implemented such an innovative approach by developing a mathematics room (see http://dmmh.no/for-barnehagene/matematikkrommet). In our next study, we will investigate what effect this approach has on students' MA.

Does MSE affect students' knowledge mostly indirectly via MA, or is there a direct effect as well? Our study revealed a reversal paradox. MSE's indirect effect via MA is positive, while its direct effect is the opposite, negative. This shows that working on a realistic MSE is important. Tsamir et al. (2014) describe a programme to increase teachers' aware-ness of their own mathematical abilities. Our findings suggest that a programme like this should be incorporated into the teacher training curriculum.

Note

1. The European Credit Transfer and Accumulation System (ECTS) is a system for credit accumulation and transfer. ECTS credits express the volume of learning based on the defined learning outcomes and their associated workload (European Commission, 2015, p. 10).

Acknowledgements

We appreciate Anne Hj. Nakken from the Norwegian Centre for Mathematics Education, Audrey Cooke from Curtin University in Perth, Western Australia, and Bob Perry, Emeritus Professor at Charles Sturt University and Director of Peridot Education Pty Ltd, Albury, Australia, supporting us with good advice, ideas, and proofreading. We express our thanks to Øystein Andersen from the Centre for Educational Measurement at the University of Oslo (CEMO) for helping with the trans-lation of the test instruments.

Disclosure statement

No potential conflict of interest was reported by the authors.

References

Alvestad, M., J.-E. Johansson, T. Moser, and F. Søbstad. 2009. "Status og utfordringer i norsk bar-nehageforskning." *Nordic Early Childhood Education Research* 2: 39–55.
Arbaugh, F., D. L. Ball, P. Grossman, D. E. Heller, and D. Monk. 2015. "Deans' Corner: Views on the State of Teacher Education in 2015." *Journal of Teacher Education* 66 (5): 435–445.
Bai, H., L. Wang, W. Pan, and M. Frey. 2009. "Measuring Mathematics Anxiety: Psychometric Analysis of a Bidimensional Affective Scale." *Journal of Instructional Psychology* 36 (3): 185–193.

Ball, D. L., M. H. Thames, and G. Phelps. 2008. "Content Knowledge for Teaching: What Makes it Special?" *Journal of Teacher Education* 59 (5): 389–407.

Bandura, A. 1977. "Self-Efficacy: Toward a Unifying Theory of Behavioral Change." *Psychological Review* 84: 191–215.

Bandura, A. 1986. *Social Foundations of Thought and Action: A Social Cognitive Theory.* Englewood Cliffs, NJ: Prentice-Hall, Inc.

Bates, A. B., N. I. Latham, and J.-A. Kim. 2011. "Linking Preservice Teachers' Mathematics Self-Efficacy and Mathematics Teaching Efficacy to Their Mathematical Performance." *School Science and Mathematics* 111 (7): 325–333.

Bates, A. B., N. I. Latham, and J.-A. Kim. 2013. Do I Have to Teach Math? Early Childhood Pre-Service Teachers' Fears of Teaching Mathematics. *Issues in the Undergraduate Mathematics Preparation of School Teachers,* 5. http://files.eric.ed.gov/fulltext/EJ1061105.pdf.

Beilock, S. L., E. A. Gunderson, G. Ramirez, and S. C. Levine. 2010. "Female Teachers' Math Anxiety Affects Girls' Math Achievement." *Proceedings of the National Academy of Sciences of the United States of America* 107 (5): 1860–1863.

Bekdemir, M. 2010. "The Pre-Service Teachers' Mathematics Anxiety Related to Depth of Negative Experiences in Mathematics Classroom While They Were Students." *Educational Studies in Mathematics* 75 (3): 311–328.

Benson, J. 1989. "Structural Components of Statistical Test Anxiety in Adults: An Exploratory Study." *Journal of Experimental Education* 57: 247–261.

Benz, C. 2012. "Maths Is Not Dangerous – Attitudes of People Working in German Kindergarten About Mathematics in Kindergarten." *European Early Childhood Education Research Journal* 20 (2): 249–261.

Bessant, K. C. 1995. "Factors Associated with Types of Mathematics Anxiety in College Students." *Journal for Research in Mathematics Education* 26 (4): 327–345.

Bjørseth, S. 2017. "Matematikk i barnehagen. En surveyundersøkelse om pedagogiske lederes kunnskap om matematisk arbeid i barnehagen." Master thesis, University of Oslo, Oslo. https://www.duo.uio.no/handle/10852/58021?show=full.

Blömeke, S., J. E. Gustafsson, and R. J. Shavelson. 2015. "Beyond Dichotomies Competence Viewed as a Continuum." *Zeitschrift für Psychologie* 223 (1): 3–13.

Blömeke, S., O. Thiel, and L. Jenßen. 2017. "Before, During and After Examination: Development of Prospective Preschool Teachers' Mathematics-Related Enjoyment and Self-Efficacy." *Scandinavian Journal of Educational Research.* doi:10.1080/00313831.2017.1402368.

Buchholtz, N., F. K. S. Leung, L. Ding, G. Kaiser, K. Park, and B. Schwarz. 2013. "Future Mathematics Teachers' Professional Knowledge of Elementary Mathematics From an Advanced Standpoint." *ZDM* 45 (1): 107–120.

Cahill, M. J., M. A. McDaniel, R. F. Frey, K. M. Hynes, M. Repice, J. Q. Zhao, and R. Trousil. 2018. "Understanding the Relationship Between Student Attitudes and Student Learning." *Physical Review Physics Education Research* 14 (1): 16.

Carey, E., F. Hill, A. Devine, and D. Szücs. 2016. "The Chicken or the Egg? The Direction of the Relationship Between Mathematics Anxiety and Mathematics Performance." *Frontiers in Psychology* 6. doi:10.3389/fpsyg.2015.01987.

Chipman, S. F., D. H. Krantz, and R. Silver. 1992. "Mathematics Anxiety and Science Careers Among Able College Women." *Psychological Science* 3 (5): 292–295.

Choy, S. 2002. *Nontraditional Undergraduates (NCES 2002–012).* Washington, DC: U.S. Government Printing Office. https://archive.org/details/ERIC_ED546117.

Coffman, D. L., and R. C. MacCallum. 2005. "Using Parcels to Convert Path Analysis Models Into Latent Variable Models." *Multivariate Behavioral Research* 40 (2): 235–259.

Cohen, J., and P. Cohen. 1975. *Applied Multiple Regression/Correlation Analysis for the Behavioral Sciences.* Hillsdale, NJ: Lawrence Erlbaum Associates.

Cooper, S. E., and D. A. G. Robinson. 1991. "The Relationship of Mathematics Self-Efficacy Beliefs to Mathematics Anxiety and Performance." *Measurement and Evaluation in Counseling and Development* 24 (1): 4–11.

Dehaene, S. 2011. *The Number Sense: How the Mind Creates Mathematics* (Revised and Updated ed.). New York: Oxford University Press.

Devlin, K. J. 2011. *Mathematics Education for a New Era: Video Games as a Medium for Learning.* Boca Raton, FL: Taylor & Francis.

Devlin, K. J. 2012. *Introduction to Mathematical Thinking.* Palo Alto, CA: Keith Devlin.

Devold, E. H. 2008. *Fem, seks – det kommer en heks – praktisk matematikk i barnehagen.* Oslo: GAN Aschehoug.

Devold, E. H. 2010. *En, to – støvel og sko.* Oslo: GAN Aschehoug.

Dowker, A., A. Sarkar, and C. Y. Looi. 2016. "Mathematics Anxiety: What Have We Learned in 60 Years?" *Frontiers in Psychology* 7. doi:10.3389/fpsyg.2016.00508.

Dunekacke, S., L. Jenßen, and S. Blömeke. 2015. "Effects of Mathematics Content Knowledge on Pre-School Teachers' Performance: A Video-Based Assessment of Perception and Planning Abilities in Informal Learning Situations." *International Journal of Science and Mathematics Education* 13 (2): 267–286.

Dunekacke, S., L. Jenßen, K. Eilerts, and S. Blömeke. 2016. "Epistemological Beliefs of Prospective Preschool Teachers and Their Relation to Knowledge, Perception, and Planning Abilities in the Field of Mathematics: a Process Model." *ZDM Mathematics Education* 48 (1-2): 125–137.

Eiser, J. R., R. H. Fazio, T. Stafford, and T. J. Prescott. 2003. "Connectionist Simulation of Attitude Learning: Asymmetries in the Acquisition of Positive and Negative Evaluations." *Personality and Social Psychology Bulletin* 29: 1221–1235.

Ernest, P. 2011. *Mathematics and Special Educational Needs: Theories of Mathematical Ability and Effective Types of Intervention with low and High Attainers in Mathematics.* Saarbrücken: LAP LAMBERT Academic Publishing.

European Commission. 2015. *ECTS Users' Guide.* Luxembourg: European Union. https://ec.europa.eu/education/sites/education/files/ects-users-guide_en.pdf.

Fauskanger, J., A. Jakobsen, R. Mosvold, and R. Bjuland. 2012. "Analysis of Psychometric Properties as Part of an Iterative Adaptation Process of MKT Items for Use in Other Countries." *ZDM* 44 (3): 387–399.

Fazio, R. H. 1995. "Attitudes as Object-Evaluation Associations: Determinants, Consequences and Correlates of Attitude Accessibility." In *Attitude Strength: Antecedents and Consequences*, edited by R. E. Petty and J. A. Krosnick, 247–282. Mahwah, NJ: Lawrence Erlbaum.

Fishbein, M., and I. Ajzen. 1975. *Belief, Attitude, Intention, and Behavior: An Introduction to Theory and Research.* Reading, MA: Addison-Wesley.

Fröhlich-Gildhoff, K., I. Nentwig-Gesemann, and S. Pietsch. 2011. *Kompetenzorientierung in der Qualifizierung Frühpädagogischer Fachkräfte. Eine Expertise der Weiterbildungsinitiative Frühpädagogische Fachkräfte (WiFF).* München: Deutsches Jugendinsitut e.V.

Fuson, K., and J. W. Hall. 1983. "The Acquisition of Early Number Word Meanings: A Conceptual Analysis and Review." In *The Development of Mathematical Thinking*, edited by H. P. Ginsburg, 50–108. New York: Academic Press.

Gelman, R., and C. R. Gallistel. 1978. *The Child's Understanding of Number.* Cambridge, MA: Harvard University Press.

Gunderson, E. A., G. Ramirez, S. L. Beilock, and S. C. Levine. 2013. "Teachers' Spatial Anxiety Relates to 1st- and 2nd-Graders' Spatial Learning." *Mind, Brain, and Education* 7 (3): 196–199.

Hackett, G., and N. E. Betz. 1989. "An Exploration of the Mathematics Self-Efficacy/Mathematics Performance Correspondence." *Journal for Research in Mathematics Education* 20 (3): 261–273.

Hancock, G. R. 2003. "Fortune Cookies, Measurement Error, and Experimental Design." *Journal of Modern Applied Statistical Methods* 2 (2): 293–305.

Hembree, R. 1990. "The Nature, Effects, and Relief of Mathematics Anxiety." *Journal for Research in Mathematics Education* 21 (1): 33–46.

Hill, F., I. C. Mammarella, A. Devine, S. Caviola, M. C. Passolunghi, and D. Szűcs. 2016. "Maths Anxiety in Primary and Secondary School Students: Gender Differences, Developmental Changes and Anxiety Specificity." *Learning and Individual Differences* 48: 45–53.

Hill, H. C., B. Rowan, and D. L. Ball. 2005. "Effects of Teachers' Mathematical Knowledge for Teaching on Student Achievement." *American Educational Research Journal* 42 (2): 371–406.

Hill, H. C., L. Sleep, J. M. Lewis, and D. L. Ball. 2007. "Assessing Teachers' Mathematical Knowledge. What Knowledge Matters and What Evidence Counts?" In *Second Handbook of Research on Mathematics Teaching and Learning*, edited by F. Lester, 111–156. Charlotte, NC: Information Age Publishing.

Ho, H. Z., D. Senturk, A. G. Lam, J. M. Zimmer, S. Hong, Y. Okamoto, S. Y. Chiu, Y. Nakazawa, and C. P. Wang. 2000. "The Affective and Cognitive Dimensions of Math Anxiety: A Cross-National Study." *Journal for Research in Mathematics Education* 31 (3): 362–379.

Hofer, B. K., and P. R. Pintrich. 2002. *Personal Epistemology: The Psychology of Beliefs About Knowledge and Knowing*. Mahwah, NJ: Lawrence Erlbaum Associates.

Hu, L. T., and P. M. Bentler. 1999. "Cutoff Criteria for Fit Indexes in Covariance Structure Analysis: Conventional Criteria Versus New Alternatives." *Structural Equation Modeling – A Multidisciplinary Journal* 6 (1): 1–55.

International Test Commission. 2005. ITC Guidelines for Translating and Adapting Tests. Accessed January 23, 2017. https://www.intestcom.org/files/guideline_test_adaptation.pdf.

Jain, S., and M. Dowson. 2009. "Mathematics Anxiety as a Function of Multidimensional Self-Regulation and Self-Efficacy." *Contemporary Educational Psychology* 34 (3): 240–249.

Jameson, M. M., and B. R. Fusco. 2014. "Math Anxiety, Math Self-Concept, and Math Self-Efficacy in Adult Learners Compared to Traditional Undergraduate Students." *Adult Education Quarterly* 64 (4): 306–322.

Jankvist, U. T., R. Mosvold, J. Fauskanger, and A. Jakobsen. 2015. "Analysing the Use of History of Mathematics Through MKT." *International Journal of Mathematical Education in Science and Technology* 46 (4): 495–507.

Jenßen, L., S. Dunekacke, W. Baack, M. Tengler, T. Koinzer, C. Schmude, H. Wedekind, M. Grassmann, and S. Blömeke. 2015a. "KomMa: Kompetenzmodellierung und Kompetenzmessung bei frühpädagogischen Fachkräften im Bereich Mathematik." In *Kompetenzerwerb an Hochschulen - Modellierung und Messung : zur Professionalisierung angehender Lehrerinnen und Lehrer sowie frühpädagogischer Fachkräfte*, edited by B. Koch-Priewe, A. Köker, J. Seifried, and E. Wuttke, 59–79. Bad Heilbrunn: Klinkhardt.

Jenßen, L., S. Dunekacke, M. Eid, and S. Blömeke. 2015b. "The Relationship of Mathematical Competence and Mathematics Anxiety an Application of Latent State-Trait Theory." *Zeitschrift Fur Psychologie – Journal of Psychology* 223 (1): 31–38.

Jenßen, L., O. Thiel, S. Dunekacke, and S. Blömeke. Forthcoming. *Relation Between Affective-Motivational and Cognitive Competence of Preservice Preschool Teachers in the Field of Mathematics. Methoden und Evaluation*. Berlin: Freie Universität Berlin.

Jerusalem, M., and L. Satow. 1999. "Schulbezogene Selbstwirksamkeitserwartung." In *Skalen zur Erfassung von Lehrer- und Schülermerkmalen*, edited by R. Schwarzer and M. Jerusalem, 15. Berlin: Freie Universität.

Karp, K. S. 1991. "Elementary School Teachers' Attitudes Toward Mathematics: The Impact on Students' Autonomous Learning Skills." *School Science and Mathematics* 91 (6): 265–270.

Krus, D. J., and S. M. Wilkinson. 1986. "Demonstration of Properties of a Suppressor Variable." *Behavior Research Methods Instruments & Computers* 18 (1): 21–24.

Larkin, K., and R. Jorgensen. 2016. "'I Hate Maths: Why Do We Need to Do Maths?' Using iPad Video Diaries to Investigate Attitudes and Emotions Towards Mathematics in Year 3 and Year 6 Students." *International Journal of Science and Mathematics Education* 14 (5): 925–944.

Lee, J., and L. Stankov. 2013. "Higher-Order Structure of Noncognitive Constructs and Prediction of PISA 2003 Mathematics Achievement." *Learning and Individual Differences* 26: 119–130.

Lewis, J. W., and L. A. Escobar. 1986. "Suppression and Enhancement in Bivariate Regression." *The Statistician* 35 (1): 17–26.

Lid, S. E., L. F. Pedersen, and M.-L. Damen. 2018. *Underviserundersøkelsen 2017. Hovedtendenser*. Oslo: Norwegian Agency for Quality Assurance in Education (NOKUT). https://www.nokut.no/studiebarometeret/underviserundersokelsen/.

Ma, X. 1999. "Meta-analysis of the Relationship Between Anxiety Toward Mathematics and Achievement in Mathematics." *Journal for Research in Mathematics Education* 30 (5): 520–540.

McMullan, M., R. Jones, and S. Lea. 2012. "Math Anxiety, Self-Efficacy, and Ability in British Undergraduate Nursing Students." *Research in Nursing & Health* 35 (2): 178–186.

Meeks, D. 1997. "Mathematics Anxiety and Community College Mathematics Course Completion." Diss., Northern Arizona University, Flagstaff.

Messick, D. M., and J. P. Van de Geer. 1981. "A Reversal Paradox." *Psychological Bulletin* 90 (3): 582–593.

Metje, N., H. L. Frank, and P. Croft. 2007. "Can't Do Maths – Understanding Students' Maths Anxiety." *Teaching Mathematics and its Applications: An International Journal of the IMA* 26 (2): 79–88.

Mosvold, R., R. Bjuland, J. Fauskanger, and A. Jakobsen. 2011. "Similar but Different – Investigating the use of MKT in a Norwegian Kindergarten Setting." In *Proceedings of the Seventh Congress of the European Society for Research in Mathematics Education*, edited by M. Pytlak, T. Rowland, and E. Swoboda, 1802–1811. Rzeszów: University of Rzeszów, Poland.

Mosvold, R., and J. Fauskanger. 2013. "Teachers' Beliefs About Mathematical Knowledge for Teaching Definitions." *International Electronic Journal of Mathematics Education* 8 (2-3): 43–61.

Mosvold, R., and J. Fauskanger. 2015. "Kartlegging av læreres kunnskap er ikke enkelt." *Acta Didactica Norge* 9 (1): 1–16. Art. 7.

Mosvold, R., J. Fauskanger, A. Jakobsen, and K. Melhus. 2009. "Translating Test Items Into Norwegian – Without Getting Lost in Translation?" *Nordic Studies in Mathematics Education* 14 (4): 101–123.

Muthén, L. K., and B. O. Muthén. 2014. *Mplus User's Guide.* 7th ed. Los Angeles, CA: Muthén & Muthén.

Nakken, A., and O. Thiel. 2014. *Matematikkens kjerne.* Bergen: Fagbokforlaget.

Neale, D. C. 1969. "The Role of Attitudes in Learning Mathematics." *The Arithmetic Teacher* 16 (8): 631–640.

Nortvedt, G. A., and T. Bulien. 2016. *Rapport Norsk matematikkråds forkunnskapstest 2015.* Oslo: Norwegian Mathematical Council. https://matematikkradet.no/rapport2015/NMRRapport2015.pdf.

Norwegian Centre for Research Data. 2018. Database for Statistics on Higher Education (DBH). Accessed February 24, 2018. http://dbh.nsd.uib.no/statistikk/rapport.action?visningId=249&visKode=false&columns=arstall!8!studkode&index=2&formel=1005&hier=studkode!9!instkode!9!modellkode!9!progkode!9!studtypekode&sti=Barnehagel%C3%A6rerutdanning&param=arstall%3D2017!8!2016!8!2015!8!2014!8!2013!9!dep_id%3D1!9!studkode%3DBLU!9!toppnivakode%3DLN!8!HN.

Norwegian Directorate for Education and Training. 2012. *Forskrift om rammeplan for barnehagelærerutdanning.* Oslo: Ministry of Education and Research. https://www.regjeringen.no/no/dokumenter/rundskriv-f-04-12/id706946/.

Norwegian Directorate for Education and Training. 2017. *Framework Plan for the Kindergartens Content and Tasks.* Oslo: Ministry of Education and Research. https://www.udir.no/globalassets/filer/barnehage/rammeplan/framework-plan-for-kindergartens2-2017.pdf.

Oppermann, E., Y. Anders, and A. Hachfeld. 2016. "The Influence of Preschool Teachers' Content Knowledge and Mathematical Ability Beliefs on Their Sensitivity to Mathematics in Children's Play." *Teaching and Teacher Education* 58: 174–184.

Østrem, S., H. Bjar, L. R. Føsker, H. D. Hogsnes, T. T. Jansen, S. Nordtømme, and K. R. Tholin. 2009. *Alle teller mer: en evaluering av hvordan Rammeplan for barnehagens innhold og oppgaver blir innført, brukt og erfart.* Tønsberg: Høgskolen i Vestfold. http://hdl.handle.net/11250/149122.

Pajares, F., and M. D. Miller. 1994. "Role of Self-Efficacy and Self-Concept Beliefs in Mathematical Problem-Solving – A Path-Analysis." *Journal of Educational Psychology* 86 (2): 193–203.

Parker, P. D., H. W. Marsh, J. Ciarrochi, S. Marshall, and A. S. Abduljabbar. 2014. "Juxtaposing Math Self-Efficacy and Self-Concept as Predictors of Long-Term Achievement Outcomes." *Educational Psychology* 34 (1): 29–48.

Piaget, J., and A. Szeminska. 1941. *La genèse du nombre chez l'enfant.* Neuchâtel: Delachaux et Niestlé.

Pólya, G. 1977. *Mathematical Methods in Science*. Washington, DC: Mathematical Association of America.

Reikerås, E. K. L., and J. Fauskanger. 2008. "Ti er Ikke ti, men fem." In *GLSM i Barnehagen*, edited by S. Kibsgaard, 128–143. Oslo: Universitetsforlaget.

Richardson, F. C., and R. M. Suinn. 1972. "The Mathematics Anxiety Rating Scale: Psychometric Data." *Journal of Counseling Psychology* 19 (6): 551–554.

Rowland, T., and K. Ruthven 2011. *Mathematical Knowledge in Teaching (Vol. 50)*. Dordrecht: Springer.

Schoenfeld, A. H. 2007. "The Complexities of Assessing Teacher Knowledge." *Measurement: Interdisciplinary Research and Perspectives* 5 (2-3): 198–204.

Schwarzer, R., and M. Jerusalem. 1995. "Generalized Self-Efficacy Scale." In *Measures in Health Psychology: A User's Portfolio. Causal and Control Beliefs*, edited by J. Weinman, S. Wright, and M. Johnston, 35–37. Windsor: Nfer-Nelson.

Slepkov, A. D., and R. C. Shiell. 2014. "Comparison of Integrated Testlet and Constructed-Response Question Formats." *Physical Review Special Topics-Physics Education Research* 10 (2): 15.

Statistics Norway. 2013. *Ferdigheter i voksenbefolkningen. Resultater fra den internasjonale undersøkelsen om lese- og tallforståelse (PIAAC)* B. Bjørkeng (Ed.). Oslo-Kongsvinger: Statistics Norway. http://www.ssb.no/utdanning/artikler-og-publikasjoner/_attachment/141211?_ts = 1416e80e8e0.

Stipek, D. 2017, November 13. *A Stanford Professor Says We Should Teach More Math in Preschool/ Interviewer: J. Anderson*. https://qz.com/1125046.

Suh, Y. 2015. "The Performance of Maximum Likelihood and Weighted Least Square Mean and Variance Adjusted Estimators in Testing Differential Item Functioning With Nonnormal Trait Distributions." *Structural Equation Modeling: A Multidisciplinary Journal* 22 (4): 568–580.

Swars, S. L., C. J. Daane, and J. Giesen. 2006. "Mathematics Anxiety and Mathematics Teacher Efficacy: What is the Relationship in Elementary Preservice Teachers?" *School Science and Mathematics* 106 (7): 306–315.

Thiel, O. 2010. "Teachers' Attitudes Towards Mathematics in Early Childhood Education." *European Early Childhood Education Research Journal* 18 (1): 105–115.

Tirosh, D., P. Tsamir, E. Levenson, and M. Tabach. 2011. "From Preschool Teachers' Professional Development to Children's Knowledge: Comparing Sets." *Journal of Mathematics Teacher Education* 14: 113–131.

Tsamir, P., D. Tirosh, E. Levenson, M. Tabach, and R. Barkai. 2014. "Developing Preschool Teachers' Knowledge of Students' Number Conceptions." *Journal of Mathematics Teacher Education* 17 (1): 61–83.

Tsamir, P., D. Tirosh, E. Levenson, M. Tabach, and R. Barkai. 2015. "Preschool Teachers' Knowledge and Self-Efficacy Needed for Teaching Geometry: Are They Related?" In *From Beliefs to Dynamic Affect Systems in Mathematics Education: Exploring a Mosaic of Relationships and Interactions*, edited by B. Pepin and B. Roesken-Winter, 319–337. Cham: Springer International Publishing.

Valle, A. M. 2008. "Hjelp! – vi blir kalehudret." In *GLSM i barnehagen*, edited by S. Kibsgaard, 36–45. Oslo: Universitetsforlaget.

van Hiele, P. M. 1986. *Structure and Insight. A Theory of Mathematics Education*. Orlando, FL: Academic Press.

Vygotsky, L. S. 1978. *Mind in Society. The Development of Higher Psychological Processes*. Cambridge, MA: Harvard University Press.

Walker, W. R., C. N. Yancu, and J. J. Skowronski. 2014. "Trait Anxiety Reduces Affective Fading for Both Positive and Negative Autobiographical Memories." *Advances in Cognitive Psychology* 10 (3): 81–89.

Appendix

Table A1. Variances (diagonal), covariances (lower triangular matrix), and correlations (upper triangular matrix) of all manifest variables.

	EXAM	MA+	MA−	MSE1	MSE2	MSE3	MSE4	AGE
EXAM	1.44	0.31	0.36	−0.02	−0.26	−0.16	−0.15	−0.16
MA+	1.65	20.35	0.70	−0.38	−0.48	−0.48	−0.47	−0.19
MA−	3.36	24.45	59.33	−0.51	−0.60	−0.50	−0.52	−0.15
MSE1	−0.02	−1.44	−3.26	0.70	0.59	0.54	0.55	−0.02
MSE2	−0.23	−1.60	−3.38	0.36	0.54	0.55	0.60	−0.01
MSE3	−0.13	−1.52	−2.69	0.32	0.29	0.49	0.57	0.05
MSE4	−0.15	−1.81	−3.40	0.39	0.38	0.34	0.74	0.04
AGE	−1.35	−5.99	−8.16	−0.09	−0.02	0.27	0.25	49.29

Table A2. Means and English translation of the items developed by Jerusalem and Satow (1999) to measure MSE.

Item	Mean
MSE1	I am sure that I can solve difficult mathematical problems as well.	1.73
MSE2	Although I have not solved mathematical problems for a long time, I am nevertheless able to solve such problems when I receive them.	2.05
MSE4	Although it is exhausting, I am able to push myself and to accomplish high achievements in mathematics.	1.71
MSE3	I know different ways to solve difficult mathematical problems.	2.21

Note: High scores indicate high self-efficacy.

Table A3. Means and factor loadings of the items of the MAS-R by Bai et al. (2009), exploratory factor analysis with maximum likelihood extraction, and Varimax rotation with Kaiser Normalisation ($\chi^2 = 112.45$, df = 64, $p < .001$).

	Item	Mean	Factor MA−	MA+
MA11	Mathematics makes me feel uneasy.	2.40	**.835**	.362
MA07	I get a sinking feeling when I try to do math problems.	2.67	**.822**	.313
MA09	Mathematics makes me feel nervous.	2.75	**.792**	.309
MA06	I worry about my ability to solve math problems.	2.74	**.757**	.373
MA04	Mind goes blank, and I am unable to think clearly when doing my math test.	2.67	**.745**	.371
MA02	I get uptight during math tests.	3.17	**.733**	.278
MA14	Mathematics makes me feel confused.	2.95	**.661**	.425
MA08	I find math challenging.	3.89	**.513**	.238
MA01	I find math interesting.	2.53	.354	**.751**
MA13	I enjoy learning with mathematics.	2.31	.236	**.639**
MA12	Math is one of my favourite subjects.	3.61	.517	.615
MA05	Math relates to my life.	2.54	.201	**.572**
MA03	I think that I will use math in the future.	1.89	.282	**.556**
MA10	I would like to take more math classes.	3.18	.258	**.554**

Note: The positive items have been scored reverse so that a high score indicates high anxiety. Factor loadings greater than .500 and with a discriminant loading of at least .200 with other factors are shown in boldface.

Table A4. English translation of the exam tasks.

Task 1 (30%)	a) What do the following concepts mean?
	• Rotational symmetry
	• One-to-one correspondence
	• Positional notation
	• Three dimensional
	The explanation can be given as a definition or a free description of the content.
	b) For each of the four concepts explain why it is important that you as an ECEC teacher know what those concepts mean.

Task 2 (45%) As a pedagogue you shall prepare a mathematics room in your kindergarten.
  a) Describe an object (e.g. furniture, device, construction) that shall be in this room. Give professional grounds why this object shall be in your mathematics room.
  b) b) Describe an activity that shows how you would work with the object that you have chosen under a). Choose by yourself how many children should join the activity and how old they should be. Give grounds why you would work in this way.

Task 3 (25%) A relation is an *equivalence relation* if and only if it is *reflexive, symmetric* and *transitive*.
  (Nakken and Thiel 2014, p. 119)
  Boy (5 years): 'I run faster than Nils!' Adult: 'Did you run a race today?' Boy: 'No, but I know it anyway'. Adult: 'How is that?' Boy: 'Well, because Lars ran a race with Nils, and Lars won. I run faster than Lars, therefore, I run faster than Nils'.
  a) In this situation, the boy uses a property of the relation 'faster than'. Which one?
  b) Is the relation 'faster than' an equivalence relation? Why, why not?
  c) c) Give an example of an equivalence relation that is important in children's daily life in pre-school.

Table A5. Variances (diagonal), covariances (lower triangular matrix), and correlations (upper triangular matrix) of exam variables.

	T11	T12	T13	T14	T2a	T2b	T3a	T3b	T3c
T11	3.14	0.13	0.37	0.39	0.03	0.26	0.20	0.16	0.08
T12	0.32	2.04	0.34	0.30	0.05	0.22	0.19	0.29	0.37
T13	1.76	1.28	7.13	0.34	0.08	0.23	0.27	0.17	0.19
T14	1.19	0.73	1.53	2.94	0.13	0.30	0.11	0.28	0.39
T2a	0.15	0.19	0.45	0.57	6.07	0.59	0.03	0.12	0.19
T2b	2.14	1.47	2.79	2.32	6.63	20.98	0.13	0.21	0.21
T3a	0.70	0.54	1.35	0.38	0.17	1.14	3.89	0.42	0.36
T3b	1.01	1.49	1.51	1.76	1.03	3.50	3.00	13.15	0.68
T3c	0.56	2.07	2.04	2.62	1.89	3.84	2.83	9.27	15.76

5 Mathematical pedagogical content knowledge in Early Childhood Education

Tales from the 'great unknown' in teacher education in Portugal

Maria Pacheco Figueiredo, Helena Gomes and Cátia Rodrigues

ABSTRACT

The study aims to explore the specificity of Mathematics pedagogical content knowledge in Early Childhood Education Pedagogy in Initial Teacher Education. It addresses the issue by characterizing student teachers' perspectives and by analyzing student teachers' knowledge mobilized in a situation of planning for teaching. The answers to a task developed by students in an Initial Teacher Education program are analyzed in terms of the mathematical knowledge and pedagogical options presented. The results contribute to the discussion in terms of (un)balance between teacher-initiated and child-led activities. The discussion deepens the importance to assert specific/particular ways of teaching in Early Childhood Education, contrasting with the more restricted view of only adult-led moments being teaching. Strong content knowledge and pedagogical content knowledge are valued because of their relevance both at the level of adults' knowledge needed to support children initiatives and plan curricular/didactic activities and at the level of the knowledge children interact within their daily environment and routine.

Introduction

In 2006, initial Early Childhood Teacher Education in Portugal changed to a Masters Degree, with the possibility of a joint qualification: Early Childhood and Primary Education. The changes in the curricular structure of the programs and the higher expectation of specific knowledge production to support the Masters Degree created the opportunity for reflection regarding the role of Subject Didactics in Early Childhood Education. Early Childhood teachers' professional knowledge is usually considered as under researched (Anders and Rossbach 2015; Genishi, Ryan, and Yarnall 2001) with pedagogical content knowledge being named the 'great unknown' (Melendez Rojas 2008) due to its neglect in research (Lee 2017). The traditional knowledge base for Early Childhood Education was heavily contested in the beginning of the century, particularly the reliance on Development Psychology together with the idea of a developmentally appropriate practice (Dahlberg, Moss, and Pence 2003; Zimiles 2000). Related to the over reliance on development theories, content knowledge is mostly neglected (Chen and McNamee 2006; Cullen 2005; Hedges and Cullen 2005; Melendez Rojas 2008; Oppermann, Anders, and Hachfeld 2016; Walsh and Farrell 2008). Studies tend to focus how children learn, instead of what they learn, and pedagogical orientations disregard the specificity of content in those decisions (Genishi, Ryan, and Yarnall 2001). Content knowledge is also normally neglected in discussions about professional practice and curricular documents.

The study aims to explore the specificity of Mathematics pedagogical content knowledge in Early Childhood Education Pedagogy in an Initial Teacher Education setting in Portugal. It addresses the issue by analyzing student teachers' mathematical Pedagogical Content Knowledge mobilized in a teaching situation in an Early Childhood Education setting.

In the School of Education of Viseu, since 2010, the supervised teaching practice in the Masters Degree has been the responsibility of a multidisciplinary team, including specialists on the level of education (Early Childhood and Primary Education) and on specific curricular or Didactic areas (Mathematics, Language, Arts, Physical Education and Natural and Social Sciences). Discussions about Didactics in Early Childhood Education have been particularly relevant for the team. Because of the particular interest of the teachers involved, this paper is about Mathematics but some of the issues are larger than that subject and can be discussed regarding teacher education, Early Childhood Education and the relationship between content knowledge and pedagogical content knowledge, as proposed by several authors (Ball, Thames, and Phelps 2008; Lee 2017; McCray and Chen 2012; Shulman 1986, 1987).

Pedagogy or Didactics and content knowledge in Early Childhood Education

Recent approaches acknowledge the relevance of content knowledge for the learning of young children both at the level of adults' knowledge needed to support children initiatives and plan curricular/didactic activities and at the level of the knowledge children interact within their daily context and routine. This is related to views on Early Childhood Pedagogy that articulate three dimensions of the actions of the teacher/practitioner: (a) learning environment or 'backstage' (including the physical space and materials, time organization, groupings, social interactions and support relationships), (b) tasks or activities presented and directed by the teacher (instruction), and (c) interactions between adult and child focusing the child's activity, during play, for example (Ministério da Educação 2016; Moyles, Adams, and Musgrove 2002; Siraj-Blatchford 2010). Adult intervention is needed to incorporate all the processes that aim to initiate or maintain learning processes and/or to be effective in achieving educational goals (Portugal and Laevers 2018; Siraj-Blatchford et al. 2002).

Claims made in the Early Childhood literature that subject knowledge is not important appear to be based on philosophical ideals and a belief that attention to subjects might lead to inappropriate Pedagogy (Nutbrown 1999), relying on 'knowledge considered essential and unchangeable' and leading to 'memorization of content and its faithful reproduction as heart of the educational activity' (Oliveira-Formosinho and Formosinho 2012, 14). Recent research suggests direct instruction can hinder both creativity and problem solving for young children (Bonawitz et al. 2011) but the distinction between what we teach and how we teach is both crucial and historical (Hamilton 1999).

The emphasis must then also be placed in the conception of knowledge at stake. Discussing how to teach means analyzing how we envision content itself. *Didaktik*, in its North-Continental tradition, has for many decades looked at content as a transformative tool instead of something stall, objective and unquestionable, monolithic (Deng 2015; Hillen, Sturm, and Willbergh 2011; Svanes and Skagen 2017; Willbergh 2016). In this sense, *Didaktik* can be useful to understand content knowledge in Early Childhood Education. For the tradition of *Didaktik*, the social and cultural world is 'subjectified': there is content to be learned, but students are to be encouraged to find their own path demanding that the teacher looks at a prospective object of learning (content) to ask what it can and should signify to the student, and how students themselves can experience this significance (Künzli 2000, cited in Hudson 2007). This notion recognizes that people with different experiences will make different sense of the same situation which leads to the importance of paying attention to the child's as well as to the teacher's perspective in Early Childhood Education (van Oers 2010) and aiming at coordination or intersubjectivity (Pramling and Pramling-Samuelsson 2011). This is achieved through meaningful interactions between adult and child in rich, discovery-oriented and content-relevant learning environments (Bateman 2015; Björklund and Barendregt 2016; Worthington and van Oers 2016).

Both the Didactics of Early Childhood Education (Pramling and Pramling-Samuelsson 2011) and the Pedagogy of Early Childhood Education (Siraj-Blatchford 2010) suggest dimensions of action and knowledge that require strong content knowledge and pedagogical content knowledge of teachers. A *Didaktik* view of the content of learning might support the articulation between intentional teaching and play-based environments, which are often perceived as incompatible (Ryan and Goffin 2008) or in opposition (Bennett 2006). Content knowledge is often seen as siding with the teaching extreme, but research shows that teachers underestimate the importance of discipline-specific subject or content knowledge, even though they use it spontaneously to add depth to children's learning (Hedges and Cullen 2005; Oppermann, Anders, and Hachfeld 2016; Trawick-Smith, Swaminathan, and Liu 2016).

Teachers knowledge for children's mathematical learning

In recent years, Early Childhood Mathematics education has seen a strong case for the intentional teaching of Mathematics to young children as appropriate and desirable when based and framed by a play-based approach (Björklund and Barendregt 2016; van Oers 2010; Worthington and van Oers 2016). For example, the National Association for the Education of Young Children and the National Council of Teachers of Mathematics (2002) propose that Mathematics education at this ages should take two forms: use of teachable moments arising in children's everyday play – where children often explore Mathematical ideas – and intentionally organizing of learning experiences that build children's understanding over time. This is in line with the three dimensions of the Pedagogy of Early Childhood previously mentioned.

It is acknowledged that teachers enhance children's Mathematics learning when they ask questions that provoke clarifications, extensions, and development of new understandings that connect to the embedded significant Mathematics learning in play, classroom routines, and learning experiences across the curriculum. The Didactic of Mathematics is important, then, because it can support teachers' focus on Mathematics in play and everyday situations to better assist children in understanding mathematical concepts (Delacour 2016). van Oers (1996) notes that the potential of play to facilitate children's Mathematical thinking depends 'largely on educators' ability to 'seize on the teaching opportunities in an adequate way' (71). Dockett and Perry (2010) note that play does not guarantee by itself significant Mathematical development, but it offers rich possibilities when teachers follow up by engaging children in reflecting on and representing the Mathematical ideas that have emerged in their play. The authors argue that this ability requires: Mathematical knowledge, understanding the nature of children's play, particularly the characteristics of play that promote Mathematical learning and thinking, and awareness of the role of adults in promoting both play and Mathematical understanding. Worthington and van Oers (2016) further highlight how children's home cultural knowledge can underpin their pretend play and inform their mathematics. Those funds of knowledge that children bring from their homes can create rich opportunities for learning as they include aspects of mathematics. Those connections between the children's activities and funds of knowledge (Worthington and van Oers 2016) and mathematical concepts contribute not only to meaningful learning but also to the stimulation of children's cultural identity (van Oers, 2010).

Description of the study

The study was developed during the second semester of the Masters Degree in Early Childhood and Primary Education in a Portuguese setting. It addresses the specificity of Mathematics pedagogical content knowledge in Early Childhood Education in student teachers' knowledge mobilized in two teaching situations for Early Childhood Education settings. The specificity is understood as a balance between a focus on the three dimensions previously mentioned.

Teacher education in Portugal underwent changes in 2006 included in the Bologna Process revision of higher education. For Early Childhood Education, the main changes were a change from Bachelors to Masters Degree and a strengthening of content knowledge (30 ECTS minimum for each curricular area – Maths, Language, Sciences and Arts) at the Bachelors' level. This is a general Basic Education program common to all teachers for 0–12 years old, followed by a Masters Degree, usually a mixed program that qualifies for Early Childhood Education (0–6 years old) and Primary Education (6–10 years old). The strengthening of the content knowledge is also present in the Masters Degree for Early Childhood and Primary Education with 5 ECTS (out of 90) for content knowledge. In the School of Education of Viseu, these 5 ECTS were grouped in one single course (Seminar) located in the second semester. In this course, the four curricular areas work together. The study was developed in this course with the 27 students enrolled. The study was presented in the beginning of the semester and informed signed consents were collected from the students.

Data collection and analysis

In the beginning of the semester, students were asked to answer a short questionnaire that focused their conceptions about Early Childhood Education Pedagogy. The questionnaire had been tested with a different class the year before. Students were asked to distribute 100 points between 3 dimensions of Early Childhood Pedagogy regarding their importance for a quality education for children. The three dimensions were presented to them before in classes: organizing the educational environment (space, time and group), presenting tasks or activities to the children (planned and conducted by the adult), and interacting with children during activities which are initiated by the children (small group time, free time, etc.). Twenty-seven students answered this individual and anonymous questionnaire.

The course presented short episodes from Early Childhood Education or Primary Education contexts focusing children's answers or discourse relevant for the curricular areas (Mathematics, Portuguese, Arts and Physical Education, and Knowledge of the World). The instructions required the working groups to analyze the episode during class time and describe how they would react to/continue the situation. This was phase 1. The next phase required students to research about the content identified in the episode and conceive teaching situations that were relevant for the contents and for the episode. Finally, they were asked to compare their reactions and suggestions from phases 1 and 2.

The Maths' Episode presented a short clip from the movie 'To be and to have' from Nicolas Philibert where a small child is in a library with the rest of his colleagues and his teacher. The teacher starts a conversation by asking 'how high can you count?'. The conversation then proceeds to explore high numbers (thousand, billion, …) when the teacher asks 'Can we keep on counting? Without stopping?'. Seven groups answered the episode for Mathematics. Phase 1 and phase 2 answers to the tasks were analyzed for six groups, since information was incomplete for the other.

The written reports from the groups were the basis for the analysis, therefore, a documentary analysis (Bowen 2009) of the data was carried out on the participants' assignments. This was followed by a content analysis on particular sections of the documents. The content analysis adopted both a deductive and an inductive approach (Amado 2014), focused on the four foci: (a) Mathematical contents/topics identified and connections between them, (b) resources suggested for the topics, (c) planned teaching strategies, (d) dimensions of pedagogy included. Each teacher/researcher analyzed the reports autonomously and the analysis was discussed in pairs, resulting in three discussions. A fourth teacher – with a background in Didactics of Mathematics – was asked to analyze the reports. The final discussion between the three teachers/researchers allowed to strengthen the coding and resolve existing discrepancies.

Results – analysis of student teachers' knowledge from the episode

Identifying Mathematical content in the episode

All the groups identified some of the Mathematical content present in the episode. In the second phase, the number of topics identified doubled from 12 to 24, considering all groups. Most of the groups showed deeper knowledge about the topics they had identified in the first phase (one exception) and added new topics, mostly related to the ones identified on phase one. For example, one group talked about numerical systems after having identified sense of number and counting in the first phase; another group included natural numbers in phase two connecting it to infinity.

Table 1. Groups that considered the Mathematical topics per phase.

Content/topic	First phase	Second phase
Counting	5 groups	5 groups (−1; +1)
Numerical relationships	–	1 group (+1)
Sense of number	4 groups	5 groups (+1)
Numerical systems	–	1 group (+1)
Large numbers	1 group	2 groups (+1)
Infinity	2 groups	3 groups (+1)
Place value	1 group	1 group
Natural numbers	–	1 group

In Table 1, the contents or topics identified by the students are listed using the labels chosen by them. As the results show, topics that contribute to the development of children's sense of number (counting, numerical relationships, for example) were identified on its own. In phase 2, some of those connections were made by the students and explicitly referred to in their text. Therefore, in phase 2 there is a more articulated analysis of the content present in the episode. Large numbers could also be considered part of number sense but the episode very clearly has that topic in its forefront, which might justify why it is listed on its own by the students. The groups that gave it attention focused examples and contexts where big numbers are used, as well as the visualization of representations of those numbers (like the calculator). The infinitude of the natural numbers set was regarded as existence of a follower to any natural number and therefore connected to the natural numbers topic.

Phase 2 showed that students deepened their knowledge of the topics and were able to connect them to the curricular framework, using both the Early Childhood Education reference documents and the Primary Education ones.

Presenting and representing Mathematical topics

In their proposals to continue the situation presented in the episode, students mostly presented formal teaching situations where the adult taught children one of the Math topics identified. All the groups suggested resources to be used in those situations. The groups proposed different resources for presenting and representing the Mathematical concepts.

In phase 1, just after viewing the short clip and discussing it in group, students suggested using the resources listed in Table 2: multibase arithmetic blocks (MABs) and bills were the ones chosen by more groups.

The suggested resources were coded 10/13 times as not adequate. This inadequacy was identified in the relationship between resource and concept, for example, suggestion using the hair or the abacus to explain the infinity to children. But it was also identified in terms

Table 2. Number of groups per resources suggested to (re)present concepts in phase 1.

	MABs	Unstructured materials	Abacus	Money/ bills	Natural elements	Calculator	Train of numbers (place value chart)
Infinity	1		1	1	1		
Classes/units	1		1				1
Big numbers				1		1	
Relationship between number and quantity	1	2		1			

of the relationship between resource and the children's age, for example, the 'train of numbers' (a place value chart) for an Early Childhood Education context.

The problematic relationship between resource and content is clear in the suggestions regarding the topic of infinity. In the proposals that suggest to present this concept to children, two different approaches were adopted: (a) the idea of always adding an element, in order to reveal that the set of natural numbers is infinite, and (b) asking children to think of counting something that is very large in quantity (grains of sand, hair, stars in the sky) to make them realize they would have to give up counting. Counting plays a large role in children's early ideas related to infinity (Evans 1983, cited in Clements and Sarama 2007). The reasoning is usually based on the idea of always being able to add one more (Gelman and Butterworth 2005, cited in Sarama and Clements 2009).

Hence, the first idea listed, to make children think of situations where one more would always be added, then seems to be a good proposal. But in the students' suggestions, the idea is associated with resources, money and MABs, that children mostly deal with as being finite. Children manipulate the MABs that exist in the box in their Early Childhood Education center and do not think of all the existing MABs in the world, and the idea that money is scarce is usually presented to children early on. Therefore, the idea of infinity as always having one more to count is restricted by the resources suggested. On the other hand, the second idea, trying to create mental images of a very large quantity (sand, stars) is not connected to counting until it is suggested by the adult to count such large quantities. So, again, there are limitations to this suggestion.

Preschoolers often have a limited understanding of infinity (Sarama and Clements 2009), so the representations chosen to represent it are essential. The first options students chose are not very potent in terms of supporting the understanding of the concept. The abacus for infinity is, on the other hand, much further to promote children's understanding since it requires a double representation – quantities/numbers to the abacus and then to counting without stopping.

In phase 2 (Table 3), the resources still present 2/13 inadequacies in terms of significance to Early Childhood Education and children. Infinity still proved a challenge and 3/13 presented limitations in terms of representing that concept – the suggestion was of counting without stopping – but these were not coded as inadequate.

Four suggestions show a better connection to Early Childhood Education by recognizing the need to connect to children's daily life and contexts. For example, suggesting that money from the children's piggy bank was used, or that the daily date was the basis to see that numbers can represent very different quantities (by choosing as many objects as the day the month and the year), or that the school's sand box was used to start the discussion about infinity.

Table 3. Number of groups per resources suggested to (re)present concepts in phase 2.

	MABs	Unstructured resources	Abacus	Money/ bills	Natural resources	Calculator
Counting				1		
Large numbers	1			1		1
Position value			1			
Relationship between number and quantity		3			1	
Numerical relationships		1				
infinity				1	2	

Results – student teacher's perspectives on Early Childhood Pedagogy

Questionnaire

Students chose 'interactions with children' (62%) and 'organization of the educational environment' (38%) when asked about what was the most important focus for the teacher. The questionnaire also suggested a choice of the least important of the three dimensions. In this case, 'planned tasks and activities' was chosen as less important by 41% of the students.

Interacting with children was awarded more than 30 points by every student, with an average of 40.6 points. The educational environment was also highly valued by the students with an average of 34.4 points. The lower average belonged to the adult/teacher proposed tasks and activities with 24.3 points.

Students teachers' answers reflect the emphasis given by several authors to the organization of the educational environment and the interactions started from the activity of the children as distinctive of this particular level of education (Ministério da Educação 2016; Oliveira-Formosinho and Formosinho 2012; Portugal and Laevers 2018; Siraj-Blatchford 2010).

Analysis of the episode

Not knowing the intention of the teacher in the episode, the students presented several possibilities of continuity that, in most cases, meant proposing a set of tasks, to work with children, which would help them to develop their sense of number (powerful idea in terms of Portuguese curricular guidelines in working with numbers). In clear contrast to the answers in the questionnaire, students mostly chose adult planned and oriented tasks to further the teaching/learning situation presented in the episode. Besides neglecting the dimensions learning environment and interactions with children, students chose very formal situations where the teacher presents resources and ideas to the children. Very few references to children's exploratory actions (Bonawitz et al. 2011) were made.

The same episode could be continued by analyzing possibilities in the child's environment, or by using their own knowledge to create situations, tasks, play milieus, etc. that were more capable of directing children's awareness towards the learning objects (Pramling and Pramling-Samuelsson 2011), instead of having the adult present it to the children.

Conclusions

The results suggest that although there is a strong content knowledge component in their Initial Teacher Education, student teachers still have a limited and disconnected knowledge of mathematical topics, at least when they are placed in everyday situations. Research suggests it's necessary to consider teachers' content knowledge in terms of attitudes and beliefs regarding the nature of the content as they impact pedagogical content knowledge (Henderson and Hudson 2011). Studies with Early Childhood teachers and educators also highlight how attitudes and beliefs regarding Mathematics are important. Benz (2012) found a connection between educators' view of mathematics and the view of mathematical teaching and learning processes. Thiel (2010) reports an interrelation between the teachers' general attitude towards Mathematics and their mathematical beliefs and knowledge. Bates, Latham, and Kim (2013) showed that Early Childhood teachers possess a wide variety of fears towards Mathematics including having a lack of confidence in their teaching ability, a lack of teaching methods, an inability to engage their students, and a lack of mathematical content knowledge. The improvement from phase 1 to phase 2 suggests student teachers' were capable of improving their knowledge through self-directed research on the topics. The study did not consider other variables due to its small scale, exploratory nature.

Henderson and Hudson (2011) claim that it has been increasingly necessary in teacher education to address the content as well as the teaching of the content, with little time to make up for any deficit in content knowledge. Deepening students' content knowledge, specifically *specialized content knowledge*, the subdomain of 'pure' content knowledge unique to the work of teaching (Ball, Thames, and Phelps 2008), is a challenge for teacher education. More creative, context-based ways of tackling that challenge are needed (Figueiredo, Gomes, and de Matos 2018; Lisenbee and Ford 2018).

When asked to continue the episode, students revealed difficulties in conceiving or choosing ways to make the concepts understandable to children. The result was a focus on adult-centered proposals, with little acknowledgement of the need to connect with children's knowledge and experiences or to promote intersubjectivity. There was no reference to play or to the organization of an environment that would promote children's activity rich in Mathematical concepts and opportunities for interactions between adults and children. The guidance by the adult was either considered important for children's learning or perceived as what the task demanded. One very big limitation of this study is, therefore, its development in a course. A discussion with the students about their suggestions and choices would have deepen the data and allow for a better understanding of the conceptions and knowledge of the students.

There has been an emphasis in Mathematics content and competencies in several countries' Early Childhood Education and, as Grieshaber's (2008) reminds us, it is time to overcome stereotypes in young children's education regarding teaching. The discussion about what teaching means in ECE cannot mean the obliteration of important principles and dimensions of Early Childhood Pedagogy. In Portugal, the more socially valued and didactically researched Primary Education is seen as 'colonizing' Early Childhood Education teaching practices (Vasconcelos 2009). In a context of increasing accountability and rising academic expectations, the educational value of play has been questioned (Bonawitz et al. 2011; Pramling-Samuelsson and Fleer 2009; Wood 2004). The lack of sensitivity to the specificity of ECE and the negligence of play in students' proposals suggests the need for a stronger emphasis on what makes teaching specific in Early Childhood and a clearer view on the transitions for Primary Education, for example in terms of Mathematics learning and teaching (Perry, MacDonald, and Gervasoni 2015).

As more countries are moving Early Childhood Education teacher education into higher education, many at Masters level, it's important to highlight the potential fragility of Early Childhood Education specificity when integrated in contexts that do not recognize or value it.

Disclosure statement

No potential conflict of interest was reported by the authors.

References

Amado, João. 2014. *Manual de Investigação Qualitativa em Educação* [Handbook of Qualitative Research in Education]. 2nd ed. Coimbra: Imprensa da Universidade de Coimbra.

Anders, Yvonne, and Hans-Günther Rossbach. 2015. "Preschool Teachers' Sensitivity to Mathematics in Children's Play: The Influence of Math-Related School Experiences, Emotional Attitudes, and Pedagogical Beliefs." *Journal of Research in Childhood Education* 29 (3): 305–322.

Ball, Deborah Loewenberg, Mark Hoover Thames, and Geoffrey Phelps. 2008. "Content Knowledge for Teaching: What Makes It Special?" *Journal of Teacher Education* 59 (5): 389–407.

Bateman, Amanda. 2015. *Conversation Analysis and Early Childhood Education: The Co-production of Knowledge and Relationships*. Directions in Ethnomethodology and Conversation Analysis. Surrey: Ashgate.

Bates, Alan B., Nancy I. Latham, and Jin-Ah Kim. 2013. "Do I Have to Teach Math? Early Childhood Pre-service Teachers' Fears of Teaching Mathematics." *Issues in the Undergraduate Mathematics Preparation of School Teachers* 5: 1–10.

Bennett, John. 2006. "New Policy Conclusions from Starting Strong II. An Update on the OECD Early Childhood Policy Reviews." *European Early Childhood Education Research Journal* 14 (2): 141–156.

Benz, Christiane. 2012. "Maths is Not Dangerous – Attitudes of People Working in German Kindergarten About Mathematics in Kindergarten." *European Early Childhood Education Research Journal* 20 (2): 249–261.

Björklund, Camilla, and Wolmet Barendregt. 2016. "Teachers' Pedagogical Mathematical Awareness in Swedish Early Childhood Education." *Scandinavian Journal of Educational Research* 60 (3): 359–377.

Bonawitz, Elizabeth, Patrick Shafto, Hyowon Gweon, Noah D. Goodman, Elizabeth Spelke, and Laura Schulz. 2011. "The Double-Edged Sword of Pedagogy: Instruction Limits Spontaneous Exploration and Discovery." *Cognition* 120: 322–330.

Bowen, Glenn A. 2009. "Document Analysis as a Qualitative Research Method." *Qualitative Research Journal* 9 (2): 27–40.

Chen, Jie-Qi, and Gillian McNamee. 2006. "Strengthening Early Childhood Teacher Preparation: Integrating Assessment, Curriculum Development, and Instructional Practice in Student Teaching." *Journal of Early Childhood Teacher Education* 27 (2): 109–128.

Clements, D. H., and J. Sarama. 2007. "Early Childhood Mathematics Learning." In *Second Handbook of Research on Mathematics Teaching and Learning*, edited by Frank K. Lester, 461–555. Charlotte, NC: Information Age.

Cullen, Joy. 2005. "Children's Knowledge, Teachers' Knowledge: Implications for Early Childhood Teacher Education." *Australian Journal of Teacher Education* 24 (2): 13–27.

Dahlberg, Gunilla, Peter Moss, and Alan Pence. 2003. *Qualidade na Educação da Primeira Infância. Perspectivas Pós-Modernas* [Quality in Early Childhood Education: Post Modern Perspectives]. Porto Alegre: Artmed Editora.

Delacour, Laurence. 2016. "Mathematics and Didactic Contract in Swedish Preschools." *European Early Childhood Education Research Journal* 24 (2): 215–228.

Deng, Zongyi. 2015. "Content, Joseph Schwab and German Didaktik." *Journal of Curriculum Studies* 47 (6): 773–786.

Dockett, Sue, and Bob Perry. 2010. "What Makes Mathematics Play?" In *Shaping the Future of Mathematics Education. Proceedings of the 33rd Annual Conference of the Mathematics Education Research Group of Australasia*, edited by L. Sparrow, B. Kissane, and C. Hurst, 715–718. Fremantle: Merga.

Figueiredo, Maria, Helena Gomes, and Isabel Aires de Matos. 2018. "Crianças, Livros e Ideias: Literatura para a Infância e Aprendizagens Significativas na Formação Inicial de Educadores de Infância." [Children, Books and Ideas: Children's Literature and Meaningful Learning in Initial Teacher Education.] In *A Formação de Educador@s e Professor@s Na UniverCidade de Évora* [The Teacher Education of TEACHERS and educators in the Uni(ver)city of Évora], edited by Maria Assunção Folque. Évora: Universidade de Évora.

Genishi, Celia, Sharon Ryan, and Mary Yarnall. 2001. "Teaching in Early Childhood Education: Understanding Practices through Research and Theory." In *Handbook of Research on Teaching*, edited by Virginia Richardson, 4th ed., 1175–1210. Washington, DC: American Educational Research Association.

Grieshaber, Susan. 2008. "Interrupting Stereotypes: Teaching and the Education of Young Children." *Early Education & Development* 19 (3): 505–518.

Hamilton, David. 1999. "The Pedagogic Paradox (or Why No Didactics in England?)." *Pedagogy, Culture & Society* 7 (1): 135–152.

Hedges, Helen, and Joy Cullen. 2005. "Subject Knowledge in Early Childhood Curriculum and Pedagogy: Beliefs and Practices." *Contemporary Issues in Early Childhood* 6 (1): 66–79.

Henderson, Sheila, and Brian Hudson. 2011. "What is Subject Content Knowledge in Mathematics? On the Implications for Student Teachers' Competence and Confidence in Teaching Mathematics." In *Developing Quality Cultures in Teacher Education: Expanding Horizons in Relation to Quality Assurance*, edited by Eve Eisenschmidt and Erika Löfström, 175–194. Tallinn: University of Tallinn.

Hillen, Stefanie, Tanja Sturm, and Ilmi Willbergh. 2011. "Introducing Didactic Perspectives to Contemporary Challenges." In *Challenges Facing Contemporary Didactics*, edited by Stefanie Hillen, Tanja Sturm, and Ilmi Willbergh, 9–23. Munique: Waxmann Verlag.

Hudson, Brian. 2007. "Comparing Different Traditions of Teaching and Learning: What Can We Learn About Teaching and Learning?" *European Educational Research Journal* 6 (2): 135–146.

Lee, Jae Eun. 2017. "Preschool Teachers' Pedagogical Content Knowledge in Mathematics." *International Journal of Early Childhood* 49 (2): 229–243. doi:10.1007/s13158-017-0189-1.

Lisenbee, Peggy S., and Carol M. Ford. 2018. "Engaging Students in Traditional and Digital Storytelling to Make Connections Between Pedagogy and Children's Experiences." *Early Childhood Education Journal* 46 (1): 129–139.

McCray, Jennifer S., and Jie-Qi Chen. 2012. "Pedagogical Content Knowledge for Preschool Mathematics: Construct Validity of a New Teacher Interview." *Journal of Research in Childhood Education* 26 (3): 291–307. doi:10.1080/02568543.2012.685123.

Melendez Rojas, R. 2008. "Pedagogical Content Knowledge in Early Childhood: A Study of Teachers' Knowledge." PhD Thesis, Loyola University of Chicago, Chicago, IL.

Ministério da Educação. 2016. *Orientações Curriculares para a Educação Pré-Escolar* [Curricular Guidelines for Early Childhood Education]. Lisboa: Ministério da Educação.

Moyles, Janet R., S. Adams, and A. Musgrove. 2002. *Study of Pedagogical Effectiveness in Early Learning*. Research Report No. 363. Department for Education and Skills. London: Queen's Printer.

National Association for the Education of Young Children and the National Council of Teachers of Mathematics. 2002. "Early Childhood Mathematics: Promoting Good Beginnings." *Young Children* 57 (4): 60–81.

Nutbrown, C. 1999. *Threads of Thinking*. 2nd ed. London: Paul Chapman.

Oliveira-Formosinho, Júlia, and João Formosinho. 2012. *Pedagogy-in-Participation: Childhood Association Educational Perspective*. Porto: Porto Editora and CESC/UM.

Oppermann, Elisa, Yvonne Anders, and Axinja Hachfeld. 2016. "The Influence of Preschool Teachers' Content Knowledge and Mathematical Ability Beliefs on Their Sensitivity to Mathematics in Children's Play." *Teaching and Teacher Education* 58: 174–184.

Perry, Bob, Amy MacDonald, and Ann Gervasoni, eds. 2015. *Mathematics and Transition to School*. Early Mathematics Learning and Development. Singapore: Springer.

Portugal, Gabriela, and Ferre Laevers. 2018. *Avaliação em Educação Pré-Escolar – Sistema de Acompanhamento das Crianças* [Evaluation in Early Childhood Education: The Children Monitoring System]. Porto: Porto Editora.

Pramling, Niklas, and Ingrid Pramling-Samuelsson. 2011. "Introduction and Frame of the Book." In *Educational Encounters: Nordic Studies in Early Childhood Didactics*, edited by Niklas Pramling and Ingrid Pramling-Samuelsson, 1–14. Dordrecht: Springer.

Pramling-Samuelsson, Ingrid, and Marilyn Fleer. 2009. *Play and Learning in Early Childhood Settings. International Perspectives*. New York: Springer.

Ryan, S., and S. Goffin. 2008. "Missing in Action: Teaching in Early Care and Education." *Early Education and Development* 19 (3): 385–395.

Sarama, J., and D. H. Clements. 2009. *Early Childhood Mathematics Education Research: Learning Trajectories for Young Children.* New York: Routledge.

Shulman, Lee S. 1986. "Those Who Understand: Knowledge Growth in Teaching." *Educational Researcher* 15 (2): 4–14.

Shulman, Lee S. 1987. "Knowledge and Teaching: Foundations of the New Reform." *Harvard Educational Review* 57 (1): 1–23.

Siraj-Blatchford, Iram. 2010. "A Focus on Pedagogy. Case Studies of Effective Practice." In *Early Childhood Matters. Evidence from the Effective Pre-school and Primary Education Project,* edited by Kathy Sylva, Edward Melhuish, Pam Sammons, Iram Siraj-Blatchford, and Brenda Taggart, 149–165. Oxon: Routledge.

Siraj-Blatchford, Iram, Kathy Sylva, S. Muttock, R. Gilden, and D. Bell. 2002. *Researching Effective Pedagogy in the Early Years.* Research Report No. 356. Department for Education and Skills. London: Queen's Printer.

Svanes, Ingvill Krogstad, and Kaare Skagen. 2017. "Connecting Feedback, Classroom Research and Didaktik Perspectives." *Journal of Curriculum Studies* 49 (3): 334–351.

Thiel, Oliver. 2010. "Teachers' Attitudes towards Mathematics in Early Childhood Education." *European Early Childhood Education Research Journal* 18 (1): 105–115.

Trawick-Smith, Jeffrey, Sudha Swaminathan, and Xing Liu. 2016. "The Relationship of Teacher–Child Play Interactions to Mathematics Learning in Preschool." *Early Child Development and Care* 186 (5): 716–733.

van Oers, B. 1996. "Are You Sure? Stimulating Mathematical Thinking During Young Children's Play." *European Early Childhood Education Research Journal* 4 (1): 71–87.

van Oers, Bert. 2010. "Emergent Mathematical Thinking in the Context of Play." *Educational Studies in Mathematics* 74 (1): 23–37.

Vasconcelos, Teresa. 2009. *A Educação de Infância no Cruzamento de Fronteiras* [Early Childhood Education in the Crossing of Borders]. Lisboa: Texto Editora.

Walsh, Kerryann, and Ann Farrell. 2008. "Identifying and Evaluating Teachers' Knowledge in Relation to Child Abuse and Neglect: A Qualitative Study with Australian Early Childhood Teachers." *Teaching and Teacher Education* 24: 585–600.

Willbergh, Ilmi. 2016. "Bringing Teaching Back in: The Norwegian NOU 'The School of the Future' in Light of the Allgemeine Didaktik Theory of Wolfgang Klafki." *Nordisk Tidsskrift for Pedagogikk Og Kritikk* 2: 111–124.

Wood, Elizabeth. 2004. "Developing a Pedagogy of Play." In *Early Childhood Education: Society and Culture,* edited by Angela Anning, Joy Cullen, and Marilyn Fleer, 19–30. London: Sage.

Worthington, Maulfry, and Bert van Oers. 2016. "Pretend Play and the Cultural Foundations of Mathematics." *European Early Childhood Education Research Journal* 24 (1): 51–66.

Zimiles, Herbert. 2000. "On Reassessing the Relevance of the Child Development Knowledge Base to Education." *Human Development* 43: 235–245.

6 The impact of the Promoting Early Number Talk project on the development of abstract representation in mathematics

Pamela Moffett (iD) and Patricia Eaton (iD)

ABSTRACT

It is widely acknowledged that the transition from informal, everyday knowledge to formal, school-taught knowledge is a critically important stage in children's mathematical development. Research highlights the importance of making connections between the different representations of mathematics and the value of children's own mathematical graphics in bridging the gap between the concrete and the abstract. The Promoting Early Number Talk project explored the effect of 'Number Talk' – a resource book to support the development of early number – on teaching and learning in five Year 1 primary school classes in Northern Ireland. This article investigates the impact of the project on the development of abstract representation in mathematics. Findings demonstrate an increased appreciation of the importance of social interaction and the potential value of children's own mathematical graphics. It is recommended that greater prominence is given to the development of children's own mathematical graphics both within curriculum policy and classroom practice.

Introduction

The transition from informal, intuitive mathematical knowledge to formal, school-taught mathematical knowledge is a critically important stage in children's mathematical development (Ginsburg 1975; Purpura, Baroody, and Lonigan 2013; Starkey, Klein, and Wakeley 2004). Informal mathematical knowledge consists of those competencies that children acquire before or outside of school; it is often developed through spontaneous but meaningful everyday situations and includes the use of unconventional and even self-invented symbols and procedures. Formal mathematical knowledge, by contrast, consists of those skills and concepts taught in school and is characterised by the use of conventional written notation and standard written algorithms. Many children experience difficulty in mathematics at school because its abstract and formal nature is very different from the intuitive and informal mathematical knowledge that they have already acquired (Donaldson 1978; Ginsburg 1977; Hiebert 1984). Whilst formalisation is essential, it presents serious challenges for learning and teaching (Hiebert 1984). This article demonstrates the potential of early number talk in addressing these challenges.

Most of mathematics, particularly school mathematics, ultimately depends on the use of written symbols (Van Oers 1996; Woodrow 1982). Many children begin school with simple problem-solving strategies, and then as they receive formal instruction, replace them with memorised rules and procedures devoid of understanding (Hiebert 1984; Hughes 1986). Indeed, children's 'overly mechanical and inflexible' behaviour on mathematical tasks has been a long-standing issue in mathematics education research (Hiebert 1988, 333). It is believed that children's learning in mathematics could be greatly enhanced by helping them to connect written symbols with their related understandings from the outset (Hiebert 1984). As Van Oers (2000, 136) emphasises, 'The efforts of the pupils to get a better grip on symbols in a meaningful way should be considered one of the core objectives of education, especially in the domain of mathematics.'

This article will discuss the difficulties children encounter with the formal code of arithmetic and highlight the importance of helping children to make connections between their own informal, intuitive understanding and the formal written symbolism of school mathematics. The potential of children's own mathematical representations in bridging the gap between the informal and the formal will also be discussed. It is argued that social interaction, particularly talk, is an important vehicle in supporting the development of knowledge and understanding of formal representation in mathematics.

Children's difficulties with the formal code of mathematics

Based on the work of Piaget (1952), it was previously assumed that children are unlikely to understand addition and subtraction until around the age of seven. However, numerous studies have since cast doubt on this claim and provide evidence of the considerable numerical competence which young children possess before they even start school (Gelman and Gallistel 1978; Hughes 1981, 1983). Hughes (1981) demonstrates that preschool children can carry out a range of simple additions and subtractions, provided these are embedded in concrete situations – either real or hypothetical – and the numbers involved are small; yet, when the same calculations are used in contexts where there is no reference to specific objects and events, very few children can answer correctly. What the young child apparently lacks is the ability to understand the formal code of arithmetic (Hughes 1981). This code contains a number of important features which present significant difficulties for young children: it is context-free and relies heavily on written symbolism.

According to Hughes (1986), children's difficulties with questions such as 'What does one and two make?' stem from their failure to understand the abstract and formal language of mathematics. 'Children need to develop links – *or ways of translating* – between this new language and their own concrete knowledge' (Hughes 1986, 51, emphasis in original). Children must also learn to connect their concrete understanding of number with the formal written symbolism of arithmetic. Whilst the use of numerals for a range of purposes is widespread within the young child's environment, the conventional operator signs for addition and subtraction are not usually encountered until children start school. Although the exact age at which this takes place varies from country to country, children's early experiences with these abstract, written symbols (+, − and =) in school tend to be rooted in 'doing sums' (Hughes 1986). Ginsburg (1977) argues that children often fail to understand the rationale for the written methods that are 'imposed on them' in school and yet they are expected to use them. Their understanding of this formal symbolism and their ability to apply it in other contexts appear to be seriously limited (Hughes 1986).

Importance of connecting different representations

According to Bruner (1966), any domain of knowledge can be represented in three forms: enactive, iconic and symbolic. Van Engen (1949) opined that the primary goal of instruction in arithmetic should be to enable children to link mathematical symbols with the concrete objects or events that they represent. It is argued that meaning in mathematics is established precisely through the formation of these links – or what Hughes (1986) refers to as translations. Such translations are critical to problem-solving (Cockcroft 1982). In order to solve mathematical problems, children must be able to operate within the formal code of mathematics and they must be able to translate fluently between formal and concrete representations of the same problem. Yet, this is a source of great difficulty for young children. Many of children's observed difficulties can be described as a failure to connect their intuitive, experience-based knowledge with the formal symbols and rules of school arithmetic (Ginsburg 1977; Hiebert 1984; Hughes 1986). 'It is as if they inhabit two worlds, each with its own rules and procedures, but with little connection between the two' (Hughes 1986, 122).

The consequences can be extremely damaging. Failure to translate competently and confidently between different representations of mathematical problems results in a dangerous gap. Although children may be able to carry out formal procedures accurately and automatically, they may lack understanding of the rationale for these procedures and fail to recognise how they can be applied in other contexts. As such, school mathematics becomes a 'series of tricks to achieve the "right" answer' with little understanding of the underlying concepts and little relevance to the real word (Edwards and Edwards 1992, 35). More concerning, children may develop faulty written methods: 'Without any concrete underpinning, isolated mistakes can become habitual errors, and a bizarre written arithmetic can easily result' (Hughes 1986, 171).

Bridging the gap between the concrete and the formal

It is worrying that many children leave school with a major gap between their informal understanding of number and their ability to manipulate the formal written symbolism of arithmetic. Helping children to form crucial links between the concrete and the formal is arguably the single most important task in early mathematics education (Hughes 1986; Purpura, Baroody, and Lonigan 2013). Yet, understanding of the steps involved in making such transitions appears to be rather limited and requires further research (National Mathematics Advisory Panel 2008). The process of 'disembedding' children's mathematical thinking (Donaldson 1978) is not without its challenges. An accepted pedagogical tradition is to ground the development of abstract thinking in concrete contexts (Bruner 1966; Piaget 1952). However, findings from empirical studies investigating the impact of concrete materials on children's learning in mathematics are inconclusive (Hall 1998). According to Hall (1998), in a typical teaching sequence, children are usually given some time to explore concrete materials before being directed to a more systematic use of the materials, leading finally to the development of a formal symbolic representation or 'target procedure'. Pape and Tchoshanov (2001) argue that this process is only successful to the degree that the concrete material procedures are analogous to procedures with the symbols and the degree to which this connection is made explicit for the learner. Furthermore, the numerical meaning is not intrinsic to the concrete materials; it has to be imposed, and this renders the materials themselves to be potential barriers to children's understanding of number (Maclellan 2001).

Shift in emphasis: a socio-cultural perspective on learning

Hughes's (1981, 1983, 1986) research constitutes a significant theoretical shift in perspective. Since young children already demonstrate a considerable degree of numerical competence at the concrete level, it is suggested that further concrete experiences may not be the most appropriate educational activity. Rather, the focus should be on introducing the formal arithmetic code in ways which recognise and build upon the informal, context-bound skills and concepts which young children already possess (Hughes 1981, 1986). According to Hughes (1986), young children are capable of understanding difficult ideas when they are presented in interesting, engaging and meaningful contexts. He contends that greater prominence should be given to children's own informal methods of calculation as well as their own invented symbolism.

It has been established that young children already possess a range of informal and often self-invented methods of calculation before coming to school. Although these strategies may be inefficient and unreliable, they are meaningful for young children and provide an important foundation on which to build. Hughes (1986) believes that it is important for teachers to recognise and value these strategies before progressing to the more powerful standard strategies. It has also been established that young children are capable of understanding and using written symbols even before they start school. Children's own invented representations are often ingenious and of considerable personal significance and should be the basis of any early work on written symbolism (Hughes 1986). Hughes (1983) recommends that greater emphasis is placed on the introduction and use of arithmetical symbols in meaningful communicative situations where arithmetical symbols can be used to make life easier, and where there is a purpose for making translations between formal symbols and concrete objects and events. Furthermore, the role of the teacher cannot be underestimated; it is often essential that the relationship between formal and concrete representations of the same problem is explicitly taught (Hughes 1986).

This marked the beginning of a movement towards an alternative conception of the use of representation within the mathematics classroom (Bobis et al. 1999; Carruthers and Worthington 2006; Maclellan 2001; Pape and Tchoshanov 2001). From a socio-cultural perspective, all higher-order functions such as learning grow out of social interactions (Cobb and Yackel 1998; Rogoff 1995, 1998). Vygotsky (1978) highlighted the importance of cultural (or symbolic) tools in assisting the learning process within socio-cultural contexts. Children are viewed as 'powerful meaning-makers' who are capable of inventing and using their own symbols to explore and represent their mathematical ideas (Carruthers and Worthington 2006, 21). It is through social activity that the child begins to abstract mathematical meaning. According to Vygotsky (1991), the higher mental faculties are internalised forms of social interaction, particularly language. Bakhtin (1981, 1986) stressed the importance of creating knowledge together through talk. Children's own intuitive and informal representations can be negotiated and gradually refined through discourse with their peers and their teacher. 'Through this interaction within problem-solving situations, knowledge of mathematical representation(s) and mathematical under-standing emerges and develops' (Pape and Tchoshanov 2001, 124).

A crucial transition stage

Carruthers and Worthington (2011) introduced the term 'children's mathematical graphics' to refer to children's 'emergent mathematics' or meaning-based symbol use in mathematics, emphasising that the graphics, the mathematical thinking and the meanings are the children's own. Such graphics include scribbles, dots, drawings, tally marks, writing, and invented and standard symbols. It is their contention that supporting and developing children's own mathematics graphics will enable them to bridge the 'bi-cul-tural' divide and become 'bi-numerate', able to translate between their own informal, intuitive representations and the more standard, abstract representations of school math-ematics (Carruthers and Worthington 2006). However, over 25 years ago, diSessa et al. (1991, 156) observed, 'how rare it is to find instruction that trusts students to create their own representations'. More recently, Terwel et al. (2009, 28–29) opined that 'although there have been positive changes in the past decades, we believe that today, diS-essa's statement holds true for many classroom practices'. It is argued that the pressures of assessments and inspections can lead to children being rushed into abstract symbolism without this crucial transitional stage (Carruthers and Worthington 2006).

Northern Ireland policy context

Carruthers and Worthington (2006) contend that whilst Hughes's (1986) research appears to have influenced curriculum developments in England his influence on practice has been sadly lacking. According to *The Independent Review of Mathematics Teaching in Early Years Settings and Primary Schools*, 'It is comparatively rare ... to find adults supporting children in making mathematical marks as part of developing their abilities to extend and organise their mathematical thinking' (DCSF 2008, 34). However, in Northern Ireland, there is little evidence of the impact of Hughes's (1986) research on even curricu-lum policy. Within the non-statutory *Curricular Guidance for Pre-School Education* (DHSS&PS, CCEA & DENI 2006), adults are encouraged to 'extend, informally, the math-ematical experiences the children have already had in their home environment' (25) and there is a strong emphasis on the development of mathematical language. Although the guidance section on 'Language Development' recommends that children 'create pictures to convey thoughts or ideas ... experiment using symbols and patterns and engage in early attempts at writing using a variety of drawing, painting and writing materials' (25), there is no explicit reference to children's own mathematical graphics in the section on 'Early Mathematical Experiences'.

The *Northern Ireland Primary Curriculum* (CCEA 2007) was revised in 2007 and sets out the minimum requirement that should be taught at each stage of primary education. Children who have reached the age of 4 on or before 1st July start primary school at the beginning of the September of that year. The Foundation Stage comprises Years 1 and 2 of compulsory education and aims to provide a smooth transition from the pre-school education phase to primary school. An important principle underpinning the Foundation Stage is that 'young children learn best when learning is interactive, practical and enjoyable' (CCEA 2007, 15). Much of children's early mathematical understanding is to be developed through playful experiences. As children progress through the Foundation Stage, they are gradually introduced to a more formal curriculum, at a pace that takes account of their age and stage of development. Teachers have considerable flexibility to interpret the Foundation Stage programmes to suit the needs, interests and abilities of the children. Although the statutory curriculum is set out in separate *Areas of Learning* (including *Mathematics and Numeracy*), teachers are expected to integrate learning across the curriculum to enable children to transfer their skills and to use and apply their understanding in other contexts.

The Foundation Stage curriculum places a strong emphasis on mathematical talk: 'As the development of mathematical language is of fundamental importance, talking about work has a high priority in the early years' (CCEA 2007, 23). Prominence is also given to concrete experience with frequent references to a wide range of practical mathematical activities. Whilst children should be enabled to 'use appropriate mathematical language and symbols' (26), there is little reference to the development of children's own informal forms of mathematical representation. More explicit guidance regarding the various forms of mathematical representation is introduced in the programme for Key Stage 1: 'Children should communicate in oral, pictorial and written form, progressing at their own pace from informal personal language to mathematical language and from personal recording to mathematical representations and symbols' (CCEA 2007, 60). Again, the development of mathematical language is emphasised: 'talking about work has a higher priority than recording in the early years' (CCEA 2007, 60). As children progress into Key Stage 2, they are expected to develop more standard forms of recording than those used earlier. With regard to pencil and paper calculation, they are 'to develop their own personal ways of recording ... compare and discuss these, and ultimately refine and practise pencil and paper methods that are agreed and understood' (CCEA 2007, 59). It is also expected that children will 'use and extend their mathematical language ... progressing from the use of informal personal language to effective use of appropriate mathematical language' (CCEA 2007, 60).

Methods

The Promoting Early Number Talk (PENT) project brought together five teachers from four schools in Northern Ireland. The project involved teachers engaging with 'Number Talk' (Casserly, Moffett, and Tiernan 2014)'– a resource book to promote early number vocabulary – over a six-month period with the aim of addressing two main research questions:

- What is the impact on teachers' professional practice?
- What is the perceived impact on children's learning?

This paper investigates the impact of the project on the development of abstract representation in mathematics and, in so doing, considers both of the research questions.

All of the teachers were working with children in the first year of primary school and were from one geographical area of Northern Ireland. All primary schools in the region were contacted by email by the local education authority and initially 10 schools responded. Five then agreed to take part in the project but one school withdrew just before the start, leaving four participating schools. Schools were initially asked to nominate one teacher to participate but one school requested two participants and this was facilitated by the project team. All of the participants were female and the teachers had a range of teaching experiences with one having taught for fewer than 5 years, two between 5 and 10 years, one between 11 and 15 years, and one more than 30 years. Four of the 5 teachers had a four-year Bachelor of Education (BEd) qualification and one had a one-year Postgraduate Certificate in Education. Three of the teachers had a position of responsibility in their schools: one was a Mathematics and Numeracy coordinator with responsibility for this subject area in all seven years of the primary school and two were Foundation Stage coordinators with responsibility for the first two years of the primary school curriculum. This sample is largely reflective of the teaching population at this level in Northern Ireland where most of the profession is female and most will have a four-year BEd qualification. The four schools from which the teachers were drawn ranged in size from 200 pupils to 450 pupils, and most teachers had a class size of approximately 30 pupils. The schools were different in terms of the socioeconomic status of pupils with a variation from 4.5% to 21% of pupils eligible for free school meals.

The teachers spent an initial day in the University College to be introduced to the resource and the aims of the project. The first data collection consisting of a group interview and a questionnaire was also completed and a timeline for the project agreed. The timeline is shown in Table 1.

Teachers were expected to carry out activities from Number Talk on a weekly basis during the six-month project and during this time to keep a reflective journal for which they were given structured guidelines. They also followed two pupils as case studies and kept records of their progress. At the midpoint of the project, the teachers came back to the University College to take part in a discussion on progress to date and to undertake a further group interview. At this point, dates were agreed for a research assistant to video two of the weekly activities by each teacher. One was a common activity chosen by the research team and the second videoed lesson was chosen by each teacher. A final day in the University College concluded the project. Teachers presented their learning and a further group interview and questionnaire were administered.

The group interview questions were kept broadly identical throughout so that comparisons could be drawn; the questions asked related to the development and key indicators of number sense as well as mathematical language development. The interviews were audio recorded to allow for transcription. The pre- and post- questionnaires were also broadly

Table 1. Timeline of project.

	Data collection			
January 2016	Online questionnaire	Group Interview		Reflective journal kept by
March 2016		Group Interview		participants throughout project
May 2016			Video recordings	
June 2016	Online questionnaire	Group Interview		

identical and questions were based on literature in this field and were designed to elicit information on the effectiveness of the project.

Ethical approval was sought and obtained from the University College and was based on British Educational Research Association (BERA 2011) principles. As the lead researcher was an author of 'Number Talk', due consideration was given to the potential conflict of interest and interviews were carried out by the second researcher. At the outset school principals were given information on the nature of the project, data collection and storage, use of data and confidentiality and options to withdraw at any stage. Once this 'gatekeeper' had given permission for the school to be involved, participants were given similar information along with further detailed guidance on participation and were also informed about the option to withdraw. In addition, parent or guardian consent was sought and obtained for classes to take part in video recording which involved providing information on the subsequent storage and use of the video recordings.

The data collected in the group interviews, on which this paper is based, were analysed independently by the two researchers and using thematic analysis, key themes identified. Thematic analysis was chosen as a method because of the qualitative nature of the data gathered and because it facilitates gaining an understanding of commonalities from a range of voices in the data. The key themes emerged from the colour coding of similar comments by each researcher. The researchers then met to compare findings and a high degree of correlation was identified in the independent analyses. After discussion and agreement of themes, further separate refinement of the themes and sub-themes took place and a final agreement reached to ensure robustness. For the purposes of reporting findings, teachers have been coded as Teacher A, Teacher B, etc., and are therefore referred to in subsequent quotations as TA, TB, TC, TD and TE. TB and TE taught in the same school.

Findings and discussion

The three main themes to emerge were:

> Impact on teacher professionalism
> Impact on teacher practice
> Perceived Impact on children's learning

Each of these had a number of sub-themes as shown in Figure 1.

On further analysis, each of these sub-themes contained further key elements. This paper reports on the main findings from the group interviews regarding the impact of Number Talk on the development of abstract representation in mathematics, that is, the progression from concrete experience to formal recording. The further sub-themes within this are now explored.

Teaching abstract concepts

All five teachers involved in the PENT project agreed that teaching early number concepts can be very challenging but claimed that the 'Number Talk' resource book had helped to support their professional practice. TA described 'Number Talk' as 'a very back to basics

Teacher professional development	Teacher practice	Children's learning
• Appreciation of the importance of mathematical language • Awareness of ideas and opportunities for promoting number talk • Knowledge for teaching mathematics • Reflection on practice	• Teacher mathematics talk • **Progression from concrete experience to formal recording** • Assessment of children's learning	• Pupil mathematics talk • Pupil attitude • Development of number sense

Figure 1. Summary network of themes and sub-themes.

approach to numeracy', explaining that 'it gives learning a more inquisitive and interactive approach, reinforcing the practical approach to learning and the value of mathematical language'. The PENT teachers acknowledged the difficulties young children encounter when learning the formal code of arithmetic (Hughes 1981) but claimed that 'Number Talk' had helped to take away the fear of introducing abstract mathematical ideas. As TD observed:

> I sort of think, oh my goodness, how will I even begin to approach that with the class, but once you match it up with an activity you can see how simple it can be. Even if it is during play … incidentally. It is not as daunting for us then … It seems so abstract for a four or five year old. Yet the guidance there gives it very simply, brings it down so that takes away the fear for us.

Importance of mathematical language

The NI curriculum places a heavy emphasis on the development of mathematical language in the early years (CCEA 2007). However, there is limited evidence to suggest that it has had a significant impact on practice. Even though the curriculum has been in place for over ten years, there was unanimous agreement that involvement in the project had renewed the focus for these teachers on the importance of early mathematical talk. According to TB:

> 'I think this whole project has taught me that if you can get that solid foundation of language in place in P1 it is all going to fall into place … I do think the other things are important but I think that language needs to be central and given more priority.' Indeed, the PENT study presents encouraging evidence of the positive impact of the 'Number Talk' resource on classroom mathematics talk. (Moffett and Eaton 2017)

Value of concrete experience

It seemed that participation in the project had also reinforced teachers' appreciation of the value of concrete experience in the development of early mathematical concepts, another key principle underpinning the Foundation Stage curriculum (CCEA 2007). As TA explained:

> Well practical, I think, is the basis for it, isn't it? If you haven't got the practical then … you can't really put children into formal, even like number stories and stuff like that and record it, unless they have had a load of practical.

The role of the adult

In addition, the PENT teachers acknowledged the centrality of the adult's role in supporting young children's mathematical development. According to TB, 'They need the practical and they need the play-based … but they also need to be taught explicitly as well.' Similarly, Gifford (2005, 2) claims that 'a laissez-faire approach to children learning maths in the "secret garden of play" does not work'. Socially constructed knowledge is not easily discovered when children are left to their own devices in play and so adult involvement is essential if children are to benefit from the learning opportunities that are available (Gifford 2005). In particular, the PENT teachers recognised their role in facilitating children's developing understanding by highlighting the connections between mathematical language and their own concrete experience (Hughes 1986):

> So it is really modelling it, making it explicit and then giving them some sort of tangible, hands-on, practical activity … Allowing them to explore that and play with it; test it and check it and work it out. Then they have some sort of connection and they can use the language and make links. It is actually giving them the language and the idea – the concept – all in one. (TD)

Use of meaningful contexts

It seemed that the PENT teachers had developed a greater appreciation of the value of real-life contexts. 'Real life experiences concerned with seeing maths as a tool are considerably more meaningful than the unrelated mathematics activities sometimes offered in school' (Edgington 1998, 182). A key feature of the 'Number Talk' resource book is the use of a range of meaningful contexts to enable children to make connections not only within mathematics itself but also between mathematics and the real world. For example, in the 'Fruit bowls' activity children are encouraged to combine two sets of fruit in order to find the total number of items. TD explained how she had used this activity during snack time to support Child B's understanding of addition:

> It just seems a bit more tangible and real for her to see it … It brought it to life for her a bit more … Yet if we sat down formally or I asked her to record it, she would probably switch off or disengage … I think for her it hit home through a real-life context.

Research evidence suggests that children struggle to connect their experience-based knowledge with the formal symbols of mathematics (Ginsburg 1977; Hiebert 1984; Hughes 1986). However, Hughes (1986) believes that young children are capable of understanding abstract ideas when they are presented in meaningful communicative situations. TC observed that the use of real-life contexts had also had a positive impact on children's engagement and understanding:

And making it real life. I think that has really stood out for me ... you could just see the difference in interest from the children in mathematics, and how they show a better understanding when it is in a meaningful context for them. I would continue to build on that and make more of my activities real life.

Introducing abstract written symbolism

All of the PENT teachers acknowledged that they had previously felt under pressure to progress children from practical activities directly to formal recording using abstract mathematical symbols. However, participation in the project had resulted in an important shift:

> Not pushing them into recording is another thing I think that has very much come out of it ... There is more practical work being done. (TA)

> I haven't done any recording during the last two weeks with my guided group. It has all been practical and play-based. (TB)

The PENT teachers recognised important benefits of not rushing into formal recording:

> ... but these children ... have started to talk back to me much, much more in number language. (TA)

> I have got so much more out of the children. They do enough recording. (TB)

This had appeared to greatly benefit the lower-ability children in particular. In discussing one such child, TE observed:

> I have noticed a big difference in how much he is focused during maths activities ... I couldn't believe how much mathematical language he actually has on-board ... In his book, it might look like he has nothing on board, but when I am doing the language and the observations and assessment of what he knows, he knows a lot more than what the [written] evidence suggests.

Despite these important benefits, some teachers were struggling with the resulting lack of written evidence. As TA admitted, 'Most of the time my children didn't record. I felt guilty ... .' It was also clear from the discussions that teachers were questioning the value of their monitoring and assessment procedures. As TD explained, 'I probably don't have as much in my books as I have in previous years but then when I think back I was just putting in worksheets for the sake of a book call in the 3rd term.' TB argued that gathering worksheets and collecting photos for a portfolio of evidence 'is not doing justice to the children'. She added, 'I can get a bit frustrated by getting tied up with needing evidence ... but if we didn't record any evidence, and didn't have any observations, actually at the end of the year those little children are the evidence.'

A new appreciation of children's own informal mathematical graphics

Carruthers and Worthington (2006) observe that the pressures of assessments and inspections can be interpreted as a need to rush children into formal recording with abstract symbolism without an important transitional and short-term stage involving children's own informal mathematical graphics. All five PENT teachers acknowledged that this had also been their experience. However, participation in the project had raised their awareness of the potential value of children's own mathematical graphics:

> I have more appreciation of the recording bit ... One time we were combining sets and I had whiteboards available. I said, 'If you wish you can show me what you have found out? ... One child just drew four little straight lines. Now, we weren't using sticks. It might have been cubes or dinosaurs. He just represented it pictorially with lines, and I probably would have, prior to this, not have appreciated that. (TC)

> I think, like you, I would have skipped ... I jumped over and moved on too quickly ... I went from, in the past, practical addition to 'OK here is the addition symbol, this is the equals symbol, and this is how we lay it out. This is the number sentence.' to actually going in and using first of all their recording. Letting them choose themselves ... I had never actually looked into that or allowed the children the opportunity to have that freedom to record it ... Then taking that and showing everybody all the different ways. That was lovely! (TD)

TD explained how the use of children's own graphics had served as a springboard for the introduction of abstract mathematical symbols: 'I valued, and showed everybody. "Look at the different ways that we can do it." It fed naturally into introducing the formal symbols ... It was using theirs as a springboard.' Similarly, Carruthers and Worthington (2006) believe that children's personal representations support them in bridging the 'bi-cultural' divide. As TA explained, 'I think there is a gradual process. You don't just jump from practical to writing number stories. There are all the different areas of recording.' Furthermore, TB acknowledged the need to be prepared to 'go back ... being fluid between the three'.

Conclusion

The literature has demonstrated very clearly the importance of helping children to connect their informal, experience-based knowledge of number with the formal symbols and procedures of school mathematics. In particular, the development of children's own mathematical graphics is considered to be a crucial transition stage in helping them to bridge the 'bi-cultural' divide, able to translate between informal and formal representations of the same problem. From a socio-cultural perspective, educational environments which value and build on children's existing knowledge and capabilities will best support and progress children's developing understanding. Social interaction is an essential feature of such enabling cultures: through participation and talk with peers and adults, children co-construct their understanding of mathematics and become full members of the mathematics community.

The PENT teachers acknowledged the challenges of introducing the formal code of arithmetic to young children. However, the 'Number Talk' resource book, with its emphasis on introducing mathematical concepts through a range of interactive, engaging and meaningful communicative situations, had helped to support and enhance their practice and they clearly believed that this had positively influenced children's understanding. In a climate where test results and inspection outcomes present significant pressures and challenges, the perceived benefits of the 'Number Talk' resource book on classroom practice appeared to outweigh concerns regarding the introduction of abstract written symbolism. There was also evidence to suggest that the shift in classroom practice was influencing teachers' beliefs about the development of mathematical understanding:

> I am not so nervous or anxious about a child now that can't record … It is far more important that the understanding is there through the language. That has been a massive learning curve for me in terms of their recordings and what the emphasis should be on. I think that that will reap benefits down the school … I have turned a real corner this year in that respect … That would be the main learning that I have had, that the practical is so important and the language. (TB)

As Guskey (2002) explains, when teachers see that a new innovation works well in their classrooms, then change their attitudes and beliefs will follow. Furthermore, teacher attitudes and beliefs impact on how mathematics teaching is purposefully incorporated in the early years (Lee and Ginsburg 2009).

There was evidence to suggest that the PENT project had helped to narrow the gap between Northern Ireland Curriculum policy, with its focus on mathematical language and practical experience, and their practice. However, it appears that there is a much wider gap – between research, policy and practice – that deserves further attention. Greater prominence must be given to the development of children's own mathematical graphics as a bridge between the informal and the formal, both within policy documentation and classroom practice.

This was a small-scale study undertaken in a single geographical region but nonetheless, the impact on practice among this group of teachers was demonstrable and indicates the potential sector-wide value of allowing teachers time and space to reflect on what works in developing mathematical understanding and engaging purposefully with literature and relevant resources such as Number Talk. As one teacher (TD) stated:

> It doesn't always have to be some kind of formal recording or something written down … All that language and practical experience that they are getting. It is so much more meaningful than doing an addition sum, a whole page, or an adding booklet or something like that. It is you giving that direct teaching and using the language and the vocabulary there. (TD)

Acknowledgements

This project was funded by internal seed-funding from Stranmillis University College, Belfast.

Disclosure statement

No potential conflict of interest was reported by the authors.

ORCID

Pamela Moffett ⓘ http://orcid.org/0000-0002-4339-6549
Patricia Eaton ⓘ http://orcid.org/0000-0002-6576-2773

References

Bakhtin, M. M. 1981. *The Dialogic Imagination: Four Essays*, edited by M. Holquist. Austin: University of Texas Press.
Bakhtin, M. M. 1986. *Speech Genre and Other Late Essays*, edited by C. Emerson and M. Holquist. Austin: University of Texas Press.

BERA (British Educational Research Association). 2011. *Ethical Guidelines for Educational Research*. Accessed February 20, 2017. https://www.bera.ac.uk/wp-content/uploads/2014/02/BERA-Ethical-Guidelines-2011.pdf.

Bobis, J., J. Mulligan, T. Lowrie, and M. Taplin. 1999. *Mathematics for Young Children: Challenging Children to Think Mathematically*. Sydney: Prentice Hall.

Bruner, J. S. 1966. *Toward a Theory of Instruction*. Cambridge: Bellcnap Press.

Carruthers, E., and M. Worthington. 2006. *Children's Mathematical Graphics: Making Marks, Making Meaning*. 2nd ed.London: SAGE.

Carruthers, E., and M. Worthington. 2011. *Understanding Children's Mathematical Graphics: Beginnings in Play*. Maidenhead: Open University Press.

Casserly, A. M., P. Moffett, and B. Tiernan. 2014. *Number Talk: A Resource to Promote Understanding and Use of Early Number Language*. ISBN Number: 978-0-9928988-09.

CCEA. 2007. *The Northern Ireland Curriculum: Primary*. Belfast: CCEA.

Cobb, P., and E. Yackel. 1998. "A Constructivist Perspective on the Culture of the Mathematics Classroom." In *The Culture of the Mathematics Classroom*, edited by F. Seeger, V. Voigt, and U. Waschescio, 159–189. Cambridge: Cambridge University Press.

Cockcroft, W. H. 1982. *Mathematics Counts*. London: HMSO.

DCSF (Department for Children, Schools and Families). 2008. *The Independent Review of Mathematics Teaching in Early Years Settings and Primary Schools. Final Report – Sir Peter Williams*. London: DCSF.

DHSS&PS, CCEA & DENI. 2006. *Curricular Guidance for Pre-school Education*. Belfast: CCEA.

DiSessa, A. A., D. Hammer, B. Sherin, and T. Kolpakowski. 1991. "Inventing Graphing: Meta-Representational Expertise in Children." *Journal of Mathematics Behaviour* 10 (2): 117–160.

Donaldson, M. 1978. *Children's Minds*. New York: Norton.

Edgington, M. 1998. *The Nursery Teacher in Action*. London: Paul Chapman.

Edwards, S., and G. Edwards. 1992. "Building Bridges Between Concrete and Abstract Conceptualization of Number in Young Children: A Strategy Using Make-believe Signifiers." *Early Child Development and Care* 82 (1): 27–36.

Gelman, R., and C. R. Gallistel. 1978. *The Child's Understanding of Number*. London: Harvard University Press.

Gifford, S. 2005. *Teaching Mathematics 3-5: Developing Learning in the Foundation Stage*. Maidenhead: Open University Press.

Ginsburg, H. P. 1975. "Young Children's Informal Knowledge of Mathematics." *Journal of Children's Mathematical Behavior* 1 (3): 63–156.

Ginsburg, H. P. 1977. *Children's Arithmetic: The Learning Process*. Oxford: Van Nostrand.

Guskey, T. R. 2002. "Professional Development and Teacher Change." *Teachers and Teaching: Theory and Practice* 8 (3/4): 381–391.

Hall, N. 1998. "Concrete Representations and the Procedural Analogy Theory." *Journal of Mathematical Behaviour* 17 (1): 33–51.

Hiebert, J. 1984. "Children's Mathematics Learning: The Struggle to Link Form and Understanding." *The Elementary School Journal* 84 (5): 497–513.

Hiebert, J. 1988. "A Theory of Developing Competence with Written Mathematical Symbols." *Education Studies in Mathematics* 19 (3): 333–355.

Hughes, M. 1981. "Can Preschool Children Add and Subtract?" *Educational Psychology* 1 (3): 207–219.

Hughes, M. 1983. "Teaching Arithmetic to Pre-school Children." *Educational Review* 35 (2): 163–173.

Hughes, M. 1986. *Children and Number: Difficulties in Learning Mathematics*. Oxford: Blackwell.

Lee, J. S., and H. P. Ginsburg. 2009. "Early Childhood Teachers' Misconceptions about Mathematics Education for Young Children in the United States." *Australasian Journal of Early Childhood* 34 (4): 37–45.

Maclellan, E. 2001. "Representing Addition and Subtraction: Learning the Formal Conventions." *European Early Childhood Education Research Journal* 9 (1): 73–86.

Moffett, P., and P. Eaton. 2017. "The Impact of the Promoting Early Number Talk Project on Classroom Mathematics Talk." *Early Child Development and Care.* doi:10.1080/03004430. 2017.1412954.

National Mathematics Advisory Panel. 2008. *Foundations for Success: The Final Report of the National Mathematics Advisory Panel.* Washington, DC: US Department of Education.

Pape, S. J., and M. A. Tchoshanov. 2001. "The Role of Representation(s) in Developing Mathematical Understanding." *Theory into Practice* 40 (2): 118–127.

Piaget, J. 1952. *The Child's Conception of Number.* London: Routledge & Kegan Paul.

Purpura, D. J., A. J. Baroody, and C. J. Lonigan. 2013. "The Transition from Informal to Formal Mathematical Knowledge: Mediation by Numeral Knowledge." *Journal of Educational Psychology* 105 (2): 453–464.

Rogoff, B. 1995. "Observing Sociocultural Activity on Three Planes: Participatory Appropriation, Guided Participation and Apprenticeship." In *Sociocultural Studies of Mind*, edited by J. Wertsch, P. Del Rio, and A. Alvarez, 139–164. New York: Cambridge University Press.

Rogoff, B. 1998. "Cognition as a Collaborative Process." In *Handbook of Child Psychology. Vol. 2: Cognition, Perception and Language*, edited by W. Damon, D. Kuhn, and R. Siegler, 679–744. New York: Wiley.

Starkey, P., A. Klein, and A. Wakeley. 2004. "Enhancing Young Children's Mathematical Knowledge through a Pre-kindergarten Mathematics Intervention." *Early Childhood Research Quarterly* 19 (1): 99–120.

Terwel, J., B. van Oers, I. van Dijk, and P. van den Eeden. 2009. "Are Representations to Be Provided or Generated in Primary Mathematics Education? Effects on Transfer." *Educational Research and Evaluation* 15 (1): 25–44.

Van Engen, H. 1949. "An Analysis of Meaning in Arithmetic." *Elementary School Journal* 49 (6): 321–329. 395–400.

Van Oers, B. 1996. "Learning Mathematics as a Meaningful Activity." In *Theories of Mathematical Learning*, edited by L. Steffe, B. Nesher, P. Cobb, G. Golden, and B. Greer, 91–114. Hillsdale, NJ: Lawrence Erlbaum Ass.

Van Oers, B. 2000. "The Appropriation of Mathematical Symbols: A Psychosemiotic Approach to Mathematical Learning." In *Symbolizing and Communicating in Mathematics Classrooms*, edited by P. Cobb, E. Yackel, and K. McClain, 133–176. London: Lawrence Erlbaum Associates.

Vygotsky, L. S. 1978. *Mind in Society: The Development of Higher Psychological Processes.* Cambridge, MA: Harvard University Press.

Vygotsky, L. 1991. "The Genesis of Higher Mental Functions." In *Learning to Think*, edited by P. Light, S. Sheldon, and B. Woodhead, 34–63. London: Routledge.

Woodrow, D. 1982. "Mathematical Symbolism." *Visible Language* 16 (3): 289–302.

7 The role of and connection between systematization and representation when young children work on a combinatorial task

Hanna Palmér and Jorryt van Bommel

ABSTRACT
This article is about the systematization and representation young children spontaneously use when they are working on a combinatorial task. In this article, documentations from 123 children working on the same task are analysed. The question asked is if there are any connections between the systematizations and representations used in the documentations and how the children solve the task. The results indicate that there are some connections between systematization and representations and that both prepossess children's solutions. In this paper, we provide some possible reasons; however, we also state that more studies are needed to give deeper insights on these issues.

Introduction

This article is about the systematization and representation 123 six-year-old children spontaneously use when they are solving a (for them) challenging combinatorial task. The task was part of an educational design research study investigating the potential of teaching young children mathematics through problem-solving. Even though problem-solving in mathematics is emphasized in the syllabus in many countries (Lesh and Zawojewski 2007) few early childhood education programmes provide mathematical challenging activities (Cross, Woods, and Schweingruber 2009; Perry and Dockett 2008). This is why we wanted to investigate the potential of teaching young children mathematics through problem-solving. However, the main issue in this article will not be on problem-solving in mathematics but on analysing children's documentations from one of the problem-solving tasks used in the study. Three questions will be elaborated on the following:

- Are there any connections between the systematization of children's documentations and how they solve the combinatorial task?

- Are there any connections between the representations the children use and how they solve the combinatorial task?
- Are there any connections between the systematization and representation used by the children?

The article is organized as follows: it starts with a presentation of the context of the study, followed by sections about combinatorics, systematization and representation. After that, the design of the study is presented followed by the analysis of the documentations. Finally, some discussions and conclusions.

The context of the study

The study was conducted in Swedish preschool class which is an optional year of schooling that children aged six can attend the year before formal schooling begins. Even though it is optional almost all Swedish six-year-olds attend preschool class (National Agency for Education 2014a). The aim of preschool class is to facilitate a smooth transition between preschool and school and prepare children for the next step of their education. To facilitate a smooth transition, preschool classes are to prepare children for the way of working in school but having their grounding in the traditions of play in preschool. Even though the preschool class was implemented in 1998, a national curriculum for preschool class was implemented first in 2017. Thus, when this study was conducted there were no national regulations regarding what mathematics content or how mathematics should be taught in preschool class. The majority of the preschool class teachers are educated as preschool teachers, but leisure time pedagogues and primary teachers are also working in preschool class (National Agency for Education 2014b).

There is a diversity (national and international) in both research and early childhood education regarding *how* early mathematics should be designed and *what* constitutes an appropriate content (Palmér and Björklund 2016). While some emphasize that the teaching of mathematics should include material, drama or pictures others have shown that such material can make it difficult for children to discern the intended mathematical objects and processes (Björklund 2014; Dowker 2005). Further, what qualifies as *young* children differ between countries and in several other countries children are much younger than the children in this study when they start formal schooling. However, the main issue in this article will not be on *how, what* or at what *age* children should be taught mathematics, but instead on exploring the systematization and representation young children spontaneously use when they first encounter a combinatorial task.

Young children and combinatorics

The question asked to the children and reported here was: In how many different ways can three toy bears sit on a sofa? The task was, therefore, an enumerative combinatorial task where the children were supposed to count the permutations for $n = 3$. Already in the late 1960s, Piaget and Inhelder investigated children working on combinatorial tasks. They described their findings in terms of cognitive development of children's combinatorial and probabilistic thinking. According to them, 2×2 or 3×3 permutation problems were not suitable for young children (age 7–8) (Piaget and Inhelder 1969).

However, English (1991, 2005) showed that a proper and meaningful context makes it possible for children to work effectively in finding permutations in combinatorial situations. She identified four principles in combinatorics for which children need to develop understanding; the principle of systematic variation, the principle of constancy, the principle of exhaustion and the principle of completion (English 1996). The principle of systematic variation meant that a different combination will occur if at least one item is varied systematically while the principle of constancy means that a different combination will occur if at least one item is kept constant while at least one other is varied systematically. The principle of exhaustion means that a constant item is exhausted when it no longer generates new combinations when the other items are varied. Finally, the principle of completion means that when all constant items have been exhausted all possible combinations have been found.

Systematization

In this article, we will analyse systematization, which will imply how the children organize their documentations when solving the combinatorial task. Four such general stages in organizing documentations, including numerical or spatial elements, have been identified: pre-structural stage, emergent stage, partial structural stage and stage of structural development (Mulligan and Mitchelmore 2009).

However, the systematization analysed in this article is not general but a goal-directed operation employed to facilitate task performance which includes both problem solution and the acquisition of domain-specific knowledge of combinatory structure (English 1996). A variety of graphic representations can be used when solving combinatorics tasks, including lists, diagrams, sketches and tables. All these may be made systematic or not and the major difficulty for young children when solving combinatorial tasks has been shown to be listing items systematically (English 2005).

Solving a combinatorial task involves making a suitable mapping of the combinations. In a study from 1991 English identified six strategies used by young children when working with combinatorial tasks: trial-and-error selection of objects – with duplicates; trial-and-error selection of objects – with rejection of duplicates; emerging pattern for the choice of objects – with rejection of duplicates; consistent and complete patterned cyclical item selection – with rejection of duplicates; emerging odometer pattern in item selection with possible item rejection and finally complete odometer pattern. These strategies are hierarchical in that the later are more effective when it comes to finding all possible combinations. In another study from 1996 English further elaborated these strategies naming them as stages; the random stage, the transitional stage and the odometer stage. In the random stage children use trial and error and checking becomes important to succeed with a task. At the transition stage children start to adopt patterns in their documentations but the pattern is not kept all through the task, instead the children revert to the trial-and error approach. This stage indicates knowledge of the principle of systematic variation and the principle of constancy mentioned in the previous section. At the odometer stage the children use an organized structure for the selection of combinations where one item is held constant while the others are varied systematically. This stage indicates knowledge also of the principle of exhaustion and principle of completion mentioned in the previous section.

When analysing children's documentations in this article we will use English notions of trial and error, transition and odometer combined with children making duplications or not.

Representations

In this article, we will also analyse the graphic representations the children use when solving the combinatorial task. Goldin and Shteingold (2001) use the terms signs, characters and objects to describe representations. These signs 'stand for (symbolize, depict, encode or represent) something other than itself' (3). Children's drawings are a first step towards using representations since they refer to objects, events, ideas and relationships beyond the surface of the drawing (Matthews 2006; Piaget and Inhelder 1969).

A diversity of words and classifications are used when describing children's representations. Most previous studies on young children's representations have been connected to quantity with few studies on young children's use of representations when solving tasks within other mathematical areas. Symbols and signs are used by Piaget and Inhelder (1969) to describe representations used by children. The difference between these two is that the symbols have some resemblance to the object they refer to. Symbols can be pictures or tally marks, invented by a child without conventions of society. Signs, on the other hand, do not resemble the objects represented but are spoken and written conventions of society, such as digits.

Children's own representations have been investigated by several researchers. Hughes (1986) focused on children's marks when representing quantity. He distinguished between idiosyncratic, pictographic, iconic and symbolic representations. Irregular representations, not related to the number of objects represented were defined as idiosyncratic ones. Children's pictures of the represented item were categorized as pictographic representations. Iconic representations are the ones based on a mark for each item whereas the standard forms like numerals or equal signs were defined as symbolic representations. Later, Carruthers and Worthington (2006) identified similar types of representations by children but added dynamic and written representations in addition to the four representations defined by Hughes.

For Heddens (1986) the connection between the concrete and abstract was of importance. Two levels were defined to describe representations used in between the concrete (objects) and the abstract (signs). At the semi-concrete level, pictures of real items, as a representation of the real situation, were considered. The semi-abstract level concerned a symbolic representation of the concrete items, with a constraint that the symbols would not look like the objects they represented.

Piaget, Hughes and Heddens all address children's representations but use different wordings. Figure 1 enlightens the similarities between for instance semi-concrete representations as described by Heddens and the pictographic representations as described by Hughes. Piaget's term symbol covers Hughes' pictographic and iconic as well as Heddens' semi-concrete and semi-abstract. Figure 1 also shows that the symbolic (Heddens) and abstract (Hughes) representations can be compared with Piaget's signs.

When analysing children's documentations in this article we will initially use Hughes (1986) notions of pictographic and iconic representations. Then, when analysing connections between the systematization and representation used by the children, also the level of abstraction (Heddens 1986; Piaget and Inhelder 1969) will be focused on.

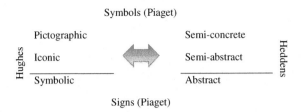

Figure 1. A composition of the various ways Piaget, Hughes and Heddens use to name and categorize representations (van Bommel and Palmér 2017).

The study

The task presented in this article is part of an educational design research study (Anderson and Shattuck 2012) on mathematical problem-solving in preschool class. There are large variations in educational design research studies, both in direction and size, but what is common is the intent to create and study new forms of instruction and to develop theories that 'guide, inform, and improve both practice and research' (Anderson and Shattuck 2012, 16). Also common in educational design research is the qualitative iterative design and its orientation to be valid for practice (Prediger, Gravemeijer, and Confrey 2015).

Ten preschool classes with a total of 123 children were involved in the study. The results in this article are based on analysis of these children's documentations from one of the problem-solving tasks used in the study. The children were verbally informed about the intervention and the interest of the researchers. The children's guardians were given written information about the study and approved their children's participation. In the written information, it was told that participation is voluntary, can be retrieved at any time, and the participants were guaranteed confidentiality. Thus, all requirements for information, approval, confidentiality and appliance advocated by the Swedish Research Council (2011) were followed. In total, only 2 children from the 10 preschools were not part of the study; however, they did participate in the lessons. The only difference was that their documentation was not collected and analysed within the study. In six of these classes, the researchers acted as teachers, whereas the actual preschool class teachers conducted the lessons in the four remaining classes.

The problem-solving task focused on in this article task was carried out with approximately 12–14 children at a time. Three plastic bears were shown to the children when introducing the task, one red, one yellow and one green. The question asked to the children was in how many different ways these toy bears could sit in a sofa. The meaning of different was explored verbally with the children. It was negotiated that different implied a movement of the bears between the three places on the sofa, not sitting behind the sofa or on its backrest, etc. Paper and colour pencils were provided, but the children did not get any instructions on how or what to document. For some time, each child worked individually on the task. When finished their individual documentation the children worked in pairs comparing their documentation to identify differences and similarities. Finally, the documentations were discussed in the whole group with a focus on how the bears had been represented, the visible systematizations and other characteristics of the documentation. Further, a discussion on the different permutations was held, where the six different combinations were displayed in a joint effort, using the plastic bears. After the whole group discussion, the children individually evaluated the difficulty of the task as well as their feelings towards the problem-solving task. This evaluation showed that the majority of the children found the task to be hard but accessible and that they had enjoyed working with it. The documentations that will be analysed in this article are the ones from the children's individual work on the task.

Analysing the data

In line with educational design research, the analysis was qualitative with the aim to enable impact and transfer of research into school practice by building theories. The first step in the analysis was to see if there were any differences in children' documentations that could be connected to the teacher of each lesson but no such patterns could be found. When analysing the documentations, we first used Hughes' (1986) pictographic and iconic representations. We found it necessary to add a category that included both these representations, as some children had used both types. Figure 2 shows examples of a pictographic and an iconic representation as well as an example of documentation with both representations.

After this, the documentations were categorized once more based on systematization. Could we see any systematizing when we looked at the permutations in each documentation? Could we, for example, see that one item had been kept invariant, that one item had been varied or did the permutations seemed to occur randomly? Thus, the analysis of systematizing in the documentations was made from an observer perspective and it is possible that children had systematizations not visible to us.

Results and analysis

Nine documentations were not possible to categorize regarding systematization as they included only one permutation or a picture of more than three bears. Table 1 shows an overview of the categorization of the remaining 114 documentations. Examples of documentations are presented in the analysis.

Only four of 114 children found 6 unique permutations when they worked individually with the task. Thus, this was a challenging task for the children. The four children who found six unique permutations used iconic representations; two with a trial-and-error approach (Figure 3, left) and two with an odometer approach (Figure 3, right). The two children who used a trial-and-error approach did not indicate knowledge of the principles

Figure 2. Examples of different representations, the first pictographic, the second iconic and the third both pictographic and iconic.

Table 1. Categorization of documentations based on representation and systematization.

		Pictographic	Pictographic/iconic	Iconic	Total	Total
Trial and error	With duplications	7	3	21	31	
	No duplications	19	2	16	37	68
Transition	With duplications	2		9	11	
	No duplications			9	9	20
Odometer	No duplications – not all solutions	7	3	14	24	
	No duplications – all solutions			2	2	26
	Total	35	8	71		114

of exhaustion or completion as the two children who used an odometer approach did (English 1996). Instead, the two children who used a trial-and-error approach had to check each of their new permutations with all their previous ones to figure out if each drawn permutation was new or not.

A total of 35 children used pictographic representations and 71 children used iconic representations. Eight children used both pictographic and iconic representations. Both the pictographic and iconic representations used by the children have a resemblance to the objects they represent (Hughes 1986). The pictographic representations are drawings of the plastic bears in the three colours. The iconic representations are drawings of different kinds but always in the colours of the toy bears. Most of the children drew dots or lines, but a few also replaced the toy bears with hearts (Figure 4).

Thus, all children used symbols (Piaget and Inhelder 1969) in their documentations. The majority of the children (71 of 114) spontaneously used an iconic representation. In studies where children document quantity, the use of symbols (pictographic and iconic representations) is often considered to represent a lower level of development than the use of signs (numerals) (Sinclair, Siegrist, and Sinclair 1983). However, in a combinatorial task, it is important to pay attention to each object per se, as well as to the relation between the objects. Therefore, iconic and pictographic representations are appropriate to use when working on combinatorial tasks.

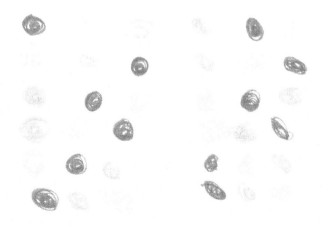

Figure 3. Examples of documentations with six duplications: the left with a trial-and-error approach, the right with an odometer approach where one item is held constant while the others are varied systematically.

Figure 4. Example of documentation with iconic representation replacing the toy bears with hearts.

Duplication of permutations was one difficulty noticed in the children's documentations (Figure 5).

Of the documentations using a trial-and-error approach or a transition approach, 30 of the iconic documentations, 3 of the combined documentations, and 9 of the pictographic documentations included duplications. However, 19 of the 28 pictographic

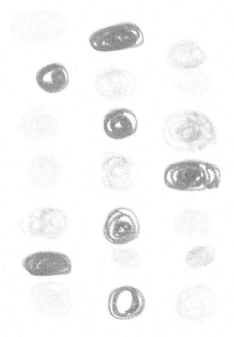

Figure 5. Example of documentation with duplications.

Figure 6. Example of documentation 'odometer – not all solutions' with each bear sitting one time at each place.

documentations using a trial-and-error approach or a transition approach consisted of unique combinations. Thus, when using a trial-and-error approach or a transition approach there was less duplication in documentations with pictographic representations (19 out of 28) than with other representations (33 out of 60) (this difference is significant on a 5% level $\chi^2 = 3.9979$; df = 1; $p < .05$). According to Heddens (1986), iconic representation is more abstract (semi-abstract) than a pictographic representation (semi-concrete). Thus, one could expect that iconic representation would imply a higher degree of success with the task and not the opposite. Time might be an important factor here. First of all, it takes a considerably longer time to draw a toy bear than to draw an iconic representation. Thus, drawing pictorial representations implies more time to think and might diminish the number of duplications. In fact, none of the documentations with pictographic representation had a solution with more than five permutations. Thus, we could conclude that children using pictographic representations made fewer permutations which, in turn, reduced the risk for duplications.

Drawing iconic representations might also lead to more duplications as the problem is not given time to be internalized, and therefore the problem 'in how many different ways can three bears sit on a sofa' may have changed into 'drawing three coloured dots on a paper'. Drawing toy bears can be experienced as more real, closer to the original problem. Some of the children could be heard saying things like 'Now it is your turn to sit in the middle' and 'He has already been in the middle'. The problem seemed to be internalized and more seldom resulted in duplications. This is in line with Devlin (2000), who states that different solutions to the 'same' mathematical task can be related to individual's perception of the task, depending on the description of the task and what it pertains to.

As mentioned, based on the iconic representation being more abstract than the pictographic representation, one could expect that iconic representation would imply a higher degree of success with the task. And, a transition approach was visible more often in iconic (18 out of 71) than in pictographic (2 out of 35) documentations (this difference is significant on a 5% level:$\chi^2 = 5.906$; df = 1; $p < .05$). Further, concerning only the odometer level, there were more iconic (16 out of 26) than combined or pictographic (3, respectively, 7 out of 26) representations. This may indicate some connections between the development of systematization and the use of abstract representation.

All but one documentation in the category 'odometer – not all solutions' consisted of exactly three combinations, each bear sitting one time at each place (Figure 6).

According to English (1996), repeated selection, as in a systematic combinatorial task, goes against the wording 'different combinations'. This is especially true for young children who often interpret 'different' as different in all aspects. Thus, they do not think that keeping one item constant and changing the others constitutes a 'different combination'; instead, when each bear has been sitting one time at each place, they think of the problem as solved.

Discussion and conclusions

In this final section we will focus on the three questions raised in the introduction to the article:

- Are there any connections between the representations the children use and how they solve the combinatorial task?
- Are there any connections between the systematization of children's documentations and how they solve the combinatorial task?
- Are there any connections between the systematization and representation used by the children?

In a previous article about children's documentation when working on a combinatorial task, only representation was focused on (Palmér and van Bommel 2017). Based on the results in that article, a question was raised regarding the systematization of the representations. English (2005) indicated that listing items systematically is a major issue for young children when solving combinatorial tasks. The results presented in this article indicate some connections between the development of systematization and the use of abstract representation. At first glimpse, it may look as though iconic representations do not generate a higher level of solution for the combinatorial task; in fact, quite the opposite, as pictographic representations result in less duplication. As long as a trial-and-error approach is used, pictographic representations seem to work best. Thus, the development of representations and systematizations seems to be somewhat synchronized. As iconic representation is considered to be more abstract than pictographic representation (Heddens 1986), it is possible that iconic representation and systematization are two expressions of child's development towards abstract thinking. However, the connections between representation and systematization found in this study may also be connected to the time it takes to draw pictographic representations or to the children's interpretation of what the task was really about.

As many as 20 of the 22 children using a transition approach (with or without duplications) and an 'odometer – all solutions' approach used iconic representation. There were, however, many children who used pictographic representations on the 'odometer – not all solutions' level. This would not be in line with the hypothesis of the development of representations and systematizations as somehow synchronized. However, the 'odometer – not all solutions' category was the one where all but one child drew exactly three permutations with each bear sitting one time at each place, which indicates an interpretation of 'different' as different in all aspects. Thus, these children may not think that keeping one item constant and changing the others ends up as a 'different combination'. Such an interpretation was visible in a previous study in other Swedish preschool classes: children were to act physically as cars when figuring out 'all different combinations' three cars could park in a garage, and when each car had been at each place once, the children were satisfied with their solution (Palmér and Ebbelind 2013).

Finally, what do the results from this study have for possible implications for policy and/or practice? As mentioned, one aim with educational design research is to enable impact and transfer of research into school practice (Anderson and Shattuck 2012). The mathematical content in the problem-solving task is fairly unusual in early mathematics and when working with tasks where the mathematical content is new and challenging for the children, communication and reasoning become important elements in the lessons. Even though only four of 114 children found the six unique permutations when they worked individually with the task, all children were engaged in the whole group discussion and the children individually evaluated the task as hard but accessible and they enjoyed working with it. The results support an education based on joint reflections on experiences and documentations from working on tasks instead of an education where children beforehand are taught systematization and which representation to use. Systematization appeared to be the core when analysing children's documentation and the results indicate that the development of representations and systematizations are somehow related: children using the transition or odometer approach used iconic representations over combined or pictographic representations. Also, systematization seems to be influenced by children's interpretation of what the task is really about (e.g. a realistic situation with toy bears or 'drawing three coloured dots on a paper'). To investigate further if iconic representations and systematization are two expressions of a child's development towards abstract thinking, we have developed a digital version of the task (van Bommel and Palmér 2017). This digital version offers semi-concrete pictorial representations together with a systematic way of saving each permutation. Thus it takes into account what seems to have led to success on the part of the study presented here. It remains to be seen, however, if such an experience in any way will influence children's representation and systematization when working on a combinatorial task.

Disclosure statement

No potential conflict of interest was reported by the authors.

References

Anderson, T., and J. Shattuck. 2012. "Design-Based Research: A Decade of Progress in Education Research?" *Educational Researcher* 41 (1): 16–25.

Björklund, C. 2014. "Less Is More – Mathematical Manipulatives in Early Childhood Education." *Early Child Development and Care* 184 (3): 469–485.

Carruthers, E., and M. Worthington. 2006. *Children's Mathematics. Making Marks, Making Meaning.* 2nd ed.London: Sage.

Cross, C. T., T. A. Woods, and H. Schweingruber. 2009. *Mathematics Learning in Early Childhood. Paths Toward Excellence and Equity.* Washington, DC: The National Academies Press.

Devlin, K. 2000. *The Math Gene: How Mathematical Thinking Evolved and Why Numbers Are Like Gossip.* New York: Basic Books.

Dowker, A. 2005. *Individual Differences in Arithmetic. Implications for Psychology, Neuroscience and Education*. New York: Psychology Press.

English, L. D. 1991. "Young Children's Combinatoric Strategies." *Educational Studies in Mathematics* 22 (5): 451–474.

English, L. D. 1996. "Children's Construction of Knowledge in Solving Novel Isomorphic Problems in Concrete and Written Form." *Journal of Mathematical Behavior* 15: 81–112.

English, L. D. 2005. "Combinatorics and the Development of Children's Combinatorial Reasoning." In *Exploring Probability in School: Challenges for Teaching and Learning*, edited by G. A. Jones, 121–141. New York: Springer.

Goldin, G., and N. Shteingold. 2001. "Systems of Representations and the Development of Mathematical Concepts." In *The Roles of Representation in School Mathematics: Yearbook*, edited by A. A. Cuoco and F. R. Curcio, 1–23. Reston, VA: National Council of Teachers of Mathematics.

Heddens, J. W. 1986. "Bridging the Gap Between the Concrete and the Abstract." *The Arithmetic Teacher* 33 (6): 14–17.

Hughes, M. 1986. *Children and Number: Difficulties in Learning Mathematics*. Oxford: Blackwell.

Lesh, R., and J. Zawojewski. 2007. "Problem Solving and Modeling." In *Second Handbook of Research on Mathematics Teaching and Learning*, edited by I. F. K. Lester (Red.), 763–799. Charlotte: National Council of Teachers of Mathematics & Information Age.

Matthews, J. 2006. "Foreword." In *Children's Mathematics: Making Marks, Making Meaning*, 2nd ed., edited by E. Carruthers and M. Worthington. London: Sage.

Mulligan, J., and M. Mitchelmore. 2009. "Awareness of Pattern and Structure in Early Mathematical Development." *Mathematics Education Research Journal* 21 (2): 33–49.

National Agency for Education. 2014a. *Descriptive Data 2013 Pre-School, School and Adult Education*. Report 399.

National Agency for Education. 2014b. *Preschool Class: Assignment, Content and Quality*. Stockholm: Swedish National Agency for Education.

Palmér, H., and C. Björklund. 2016. "Different Perspectives on Possible – Desirable – Plausible Mathematics Learning in Preschool." *Nordic Studies in Mathematics Education* 21 (4): 177–191.

Palmér, H., and A. Ebbelind. 2013. "What Is Possible to Learn? Using IPads in Teaching Mathematics in Preschool." In *Proceedings of the 37th Conference of the International Group for the Psychology of Mathematics Education. Mathematics Learning Across the Life Span*, edited by A. M. Lindmeier and A. Heinze, 425–432. Kiel: PME.

Palmér, H., and van Bommel, J. (2017). Exploring the Role of Representations When Young Children Solve a Combinatorial Task. In *ICT in Mathematics Education: The Future and the Realities: Proceedings of MADIF 10*
The Tenth Research Seminar of the Swedish Society for Research in Mathematics Education, edited by J. Häggström, E. Norén, J. van Bommel, J. Sayers, O. Helenius, and Y. Liljekvist, January 26–27, 2016. Karlstad: Svensk förening för MatematikDidaktisk Forskning – SMDF.

Perry, B., and S. Dockett. 2008. "Young Children's Access to Powerful Mathematical Ideas." In *Handbook of International Research in Mathematics Education*, edited by I. L. D. English, M. B. Bussi, G. A. Jones, R. A. Lesh, B. Sriraman, and D. Tirosh (Red.), 75–108. London: Routhledge.

Piaget, J., and B. Inhelder. 1969. *The Psychology of the Child*. New York: Basic Books.

Prediger, S., K. Gravemeijer, and J. Confrey. 2015. "Design Research with a Focus on Learning Processes: An Overview on Achievements and Challenges." *ZDM Mathematics Education* 47: 877–891.

Sinclair, A., F. Siegrist, and H. Sinclair. 1983. "Young Children's Ideas About the Written Number System." In *The Acquisition of Symbolic Skills*, edited by D. R. Rogers and J. A. Sloboda, 535–542. New York: Plenum Press.

Swedish Research Council. 2011. *God Forskningssed* [Good Custom in Research].Stockholm: Vetenskapsrådet.

van Bommel, J., and H. Palmér. 2017. "Slow Down You Move Too Fast." Poster presented at CERME10, Dublin, Ireland.

8 What makes a task a problem in early childhood education?

Rafael Ramírez-Uclés, Elena Castro-Rodríguez, Juan Luis Piñeiro and Juan F. Ruiz-Hidalgo

ABSTRACT

This article begins with a theoretical discussion of the characteristics that a task should feature to be regarded as a mathematics problem suitable for pre-primary students. Those considerations are followed by a report of a classroom experience in which three problems involving quotative or partitive division were posed to pre-primary school children to determine the presence of otherwise of the respective characteristics. The findings show that the characteristics of pre-primary education problems depend on two factors: mathematical activity that engages children and a structure that favours both their understanding of the problem and the application and verification of the solutions.

1. Introduction

Pre-primary education has become a growing concern in today's societies due to the impact of quality early education on the development of civic attitudes (OECD 2016). As part of that development, early childhood mathematics education is a subject of interest for the scientific community. Research on pre-primary school children's aptitudes (Clements and Sarama 2007; Mulligan and Vergnaud 2006; Schoenfeld and Stipek 2011) has prompted a number of mathematics teachers' organisations and groups to take a position on early childhood mathematics education (NAEYC & NCTM 2010).

The significance of early mathematics learning is not associated with quality instruction, however (Clements and Sarama 2013). Problem-solving, for instance, is not included as a process to be developed in early childhood education, for it is regarded as too complex for pre-primaries. Most research on the subject has been conducted with primary, secondary school or higher education students (Lesh and Zawojewski 2007), with very few, albeit promising, studies on early childhood. The findings of some of these studies show that suitable selection and use of problems and the related solving processes encourage skill development and help teachers gain insight into their children's thought processes (Charlesworth and Leali 2012; De Castro and Hernández 2014; Matalliotaki 2012).

Given that choosing classroom mathematics problems is no easy task, particularly for such young children, this study focused on the characteristics suitable problems should feature. The definition of such characteristics is first addressed by posing the question: what elements should characterise a pre-primary task for it to constitute a problem? A review of the literature reveals the consensuses reached from different theoretical postulates. A classroom experience is described, in which the characteristics defined were empirically contrasted by posing three problems to 5-year-old children. The three multiplicative structure problems studied were drawn from earlier research that identified problem-solving aptitudes in children at that age (Davis and Pepper 1992; Nelson and Kirkpatrick 1975).

2. Theoretical framework

A theoretical discussion of the characteristics of effective tasks and problems follows.

2.1. Classroom mathematics problems in pre-primary education

As in all other stages of schooling, in pre-primary education problem-solving is a key means of developing children's mathematical knowledge (Britz and Richard 1992; Castro and Castro 2016). Solving meaningful problems contributes to the development of higher thought processes and the discovery of a series of strategies that further children's ability to solve new problems (Pólya 1945; Schoenfeld 1985). Children acquire a sense of mathematical ideas by actively participating in the solution of a variety of mathematical problems (Britz and Richard 1992).

Problems have been characterised from a number of perspectives: educational (Kilpatrick 1980), philosophical (Agre 1982) and psychological (Mayer and Wittrock 2006). One widely accepted definition describes problems as situations that involve a subject in a series of cognitive and non-cognitive, non-predetermined processes (Castro and Castro 2016; NCTM 2000; Reys et al. 2009; Van de Walle 2003). For what one solver may be a problem, then, for another may be no more than a routine exercise for which there is an immediate answer. In early childhood, problem solver's consideration has greater emphasis in light of factors such as each child's cognitive development or the greater or lesser ex-ante stimulus received.

Facing challenges and consequently solving problems comes naturally to such young children (Britz and Richard 1992). The world is new for them and they are innately curious and flexible when confronting situations for the first time (NCTM 2000). Teachers should respect and stimulate that innate problem-solving inclination based on intuitive and informal mathematical knowledge with a view to expanding and consolidating such willingness.

Against that backdrop, the question that might be posed is: what characteristics should a task feature to constitute a problem in pre-primary education?

2.2. Characterising problems in pre-primary education

Despite the establishment of a general consensus, the question of what constitutes a problem is the object of constant evolution and revision. In particular, authors such as Nelson and Kirkpatrick (1975), Van de Walle (2003), Yee (2009) and Lesh et al. (2013) have proposed lists of characteristics that an effective pre-primary task or problem should feature.

Nelson and Kirkpatrick (1975) listed seven characteristics, the first three of which stress the role of the situation. Their list includes: (a) mathematics significance, (b) the involvement of real objects, (c) the engagement of children's interest, (d) The role of action, (e) different levels of solutions, (f) the variety of physical embodiments, and (g) the possibility that children know when a problem has a solution.

Van de Walle (2003), in turn, identified characteristics of a problematic task.

- What is problematic must be the mathematics. The task must focus children's attention on the mathematical ideas implied. Their interest must be sought not only through problems, but with the mathematics used.
- Tasks must be accessible to students. The degree of difficulty must be such that it affords opportunities to build learning sequences but should not entail inaccessible challenges. That calls for good diagnostics, for given the breadth of classroom variability the literature can provide no more than guidelines.
- Tasks must require children to justify and explain their answers and procedures.
- Tasks must include clear expectations on how ideas and the solution will be shared. The use of different formats must be explained through the use of a variety of representations (drawings, words and symbols).

Yee (2009) identified four non-elementary cognitive processes that must be required of good problem-solving tasks.

- Reasoning must be complex and non-algorithmic.
- Analysis of what needs to be done should be fostered and the use of heuristic strategies incentivised.
- Mathematical concepts, processes or relationships must be explored.
- The context must be understood to arouse interest and motivate children to seek a solution.

Along these lines, Lesh et al. (2013) proposed two questions to test for problem suitability.

(a) Do the children try to make sense of the problem using their own 'real life' experiences – instead of simply trying to do what they believe that some authority (such as the teacher) considers to be correct (even if it doesn't make sense to them)? (b) When the children are aware of several different ways of thinking about a given problem, are they themselves able to assess the strengths and weaknesses of these alternatives – without asking their teacher or some other authority? (38)

The approach proposed here, which aims to identify tasks with real life, focuses on four characteristics.

- The result is not just a 'short answer'.
- Solvers must know who needs the result and why.
- Reaching the answer is a multi-stage process.
- The answer involves integrating ideas and procedures from several areas.

The aforementioned characterisations and others to be found in the literature concur in a number of significant points. One is the emphasis on children and their context, the wealth of mathematical ideas involved or the language used to understand and express solving procedures. By way of synthesis, in this study, a pre-primary classroom problem is defined as one that features the following characteristics.

(1) C1. Reasoning. The problem must explore and develop mathematical ideas through reasoning, the use of strategies, as well as a number of trial and error cycles, rather than through algorithms.
(2) C2. Contexts. The problem must refer to situations familiar to the child. Situations need not necessarily be real from an adult's standpoint: stories, films, cartoon series are also acceptable.
(3) C3. Challenge. The task must induce the child to seek the solution. That may be furthered with different representations (verbal, physical and graphic), requiring the child to handle, transform or modify materials.
(4) C4. Multiple solutions. The problem must afford different levels of solutions, which must not consist in mere short answers.
(5) C5. Expandability. The mathematical structure must be applicable to a number of situations to enable children to generalise.
(6) C6. Comprehensibility. The problem must be understandable for all children, who must be convinced that they can solve it and know when they have found the solution.

A classroom experience was conducted to validate the aforementioned characteristics. To that end, three multiplicative word problems used in earlier research (Davis and Pepper 1992; Nelson and Kirkpatrick 1975) were selected and posed to several groups of children for analysis on the grounds of the characteristics proposed.

3. Method

This qualitative-descriptive study involved exploratory research in mathematics instruction. The methodology deployed, the population studied, the research design and the problems used are described in the sections below.

The entire process was developed under a strict supervision from the children's teacher as well as the school principal. Both teacher and parents were adequately informed. Moreover, all the interactions between researchers and children were mediated by the teacher. Finally, The researchers guarantee the confidentiality of both the data collected and the results obtained.

3.1. Subjects

The subjects were 26 children, 15 girls and 11 boys, enrolled in the same class of third year, the second level early childhood (pre-primary) education at a school in Granada, Spain, were all 5–6 years old. Whilst the participants were used to handling classroom manipulatives, they had not been taught partitive or quotative division. In Spain, and hence in Andalusia, problem-solving is embedded in the curriculum for the stage 3–6 years. Some statements about this topic can be found in the preface as well as in assessment criteria. For example 'to observe the capability to solve easy mathematics problems in dai-lylife' (MEC 2016).

3.2. Problems

Multiplicative structure and more specifically quotative and partitive division, problems that had been used in earlier studies (Davis and Pepper 1992; Nelson and Kirkpatrick 1975) were chosen, and research has yielded promising results in connection with the ability of children of these ages to solve such problems (Matalliotaki 2012). More specifically, three problems consisting in two exercises each were used.

Problem 1. The pirate panda activity (Davis and Pepper 1992). The children were gathered around a table on which three plastic figurines representing pirates were set. They were given 12 coins and told: 'three pirates want to share their booty equally. How many coins does each pirate get? Help them share' (Pirate1).

In the second exercise, a fourth pirate was set on the table and the children were told that he was entitled to the same number of coins as his buddies. They were asked: 'How many coins does each pirate get? Help them share' (Pirate2) (Figure 1).

Problem 2. Loading and unloading (Nelson and Kirkpatrick 1975, 83). Here the children worked with a working board, four buildings with unfinished roofs, a lorry and 12 square counters symbolising roof tiles. They were told that the lorry was to deliver the tiles to finish the roofs and asked how many tiles it had to deliver to each building (Loading1).

In the second exercise, the manipulatives were eight buildings with unfinished roofs, a lorry and tiles. The explanation was that the lorry had to deliver the tiles to fix the roofs and that each building needed three tiles. The questions they had to answer were: 'do you think there are enough tiles for them all? How many buildings will get enough tiles to fix the roof? How many won't get any?' (Loading2) (Figure 2).

Problem 3. The ferry (Nelson and Kirkpatrick 1975, 73). Here the working board depicted a river, with a boat and 12 cars as manipulatives. The children were told that

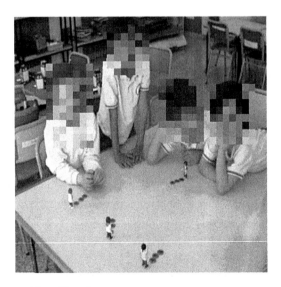

Figure 1. Team solving problem Pirate2.

Figure 2. Team solving problem Loading1.

the boat would sail several times from one shore to the other, carrying three cars each time and asked: 'how many times will the boat have to cross the river to get all the cars on the other shore? Help the boat move all the cars' (Ferry1).

Using the same manipulatives, in the second exercise the situation described was as follows. 'The boat sails from one shore to the other four times, always carrying the same number of cars. How many cars will it carry each time? Help the boat move all the cars' (Ferry2).

Materials were prepared as necessary to be used by the children to solve the problems (Figure 3).

The problems were chosen on the grounds of their conformity with the theoretical characteristics defined earlier (see Table 1).

3.3. Procedure and data collection

The experience was conducted in the middle of the academic year. Subjects were assigned by their teacher to teams of four children each, in keeping with normal classroom dynamics. The teacher took each team separately to another classroom for the experience

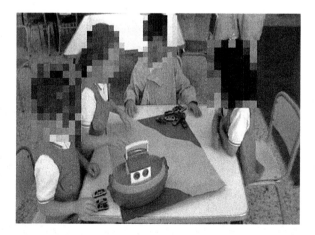

Figure 3. Team solving the Ferry problem.

Table 1. Conformity of the problems analysed to the theory on the characteristics of suitable problems.

C1	Accommodation of several solving strategies: trial and error in each phase
C2	Contextualisation: pirate figurines and coins (problem 1), mock-up and lorries (problem 2) and working board, boat and cars (problem 3)
C3	Oral and physical representation: use of manipulatives
C4	Multiple solutions: physical distribution of objects, with several levels of solution, from distribution in no order and subsequent ordering to a predetermined strategy using division
C5	Expandability: second exercises with other amounts to be distributed (with non-exact division) and divisors; generalisation to child's real-life situations involving distribution with quotative and partitive division
C6	Comprehensibility: adaptation of problem-wording to children's language and verification of the solution through physical handling of objects

and explained the problems orally, using the respective manipulatives: figurines, trucks, tiles. She stood by the children as they worked and praised their performance. Problem 1 was posed to all teams, whilst some teams did only the first or second exercise in problems 2 and 3, further to the teacher's observations about the time devoted to each.

The sessions were video recorded using one fixed and one moving camera, the fixed to obtain an overview of each team and the moving to record details of the children's actions.

3.4. Data analysis

The categories for the deductive content analysis conducted (McMillan and Schumacher 2010) were the six characteristics of pre-elementary classroom problems defined above: reasoning (C1), context (C2), challenge (C3), multiple solutions (C4), expandability (C5), and comprehensibility (C6).

The units of analysis were each team's replies to and reactions in each problem, based on both the videos and their transcriptions. The problem-solving strategies proposed by Carpenter et al. (1999) and Davis and Pepper (1992), synthesised in Tables 2 and 3, were included as subcategories under the first category, mathematical reasoning. Analyses were conducted by two of the authors independently and the partial results were subsequently harmonised by all the authors.

4. Results

Team performance in connection with the problems is described below under the categories/characteristics established as requisites for pre-primary classroom problems.

Table 2. Partitive division strategies.

Modelling strategies	S1	Count all the objects and distribute them one-by-one until none is left. Count the number of objects allocated to one of the resulting groups.
	S2	Count all the objects and distribute them two-by-two (or three-by-three ...) until none is left. Count the objects resulting from the allocation.
	S3	Begin to distribute the objects without counting the total, tallying the number allocated while some are still left and then distributing the remainder.
	S4	Count the total number of objects, allocate seven (more than appropriate) to one group and then make the necessary adjustments until the objects are equally distributed.
	S5	Divide the set of objects into equal subsets, allocating one subset to each group.
Counting strategies	S6	Skip count (3, 6, 9 ... 4, 8, 12), raising a finger or allocating one object with each number. If the last number called concurs with the number of objects to be distributed, the solution is the number of raised fingers.
Addition and subtraction strategies	SO	Find the solution by adding or subtracting.

Table 3. Quotative division strategies.

Modelling strategies	S7	Count the total number of objects needed, create groups or group the objects (five-by-five, for instance) and count the number of groups.
	S8	Group the objects five-by-five; after creating several groups, count the total number of objects allocated and use the rest to continue the grouping process.
Counting strategies	S9	Skip count: 3, 6, 9 … and use fingers, for instance, or the objects, to represent each number. Count one-by-one where necessary (3, 6, (7, 8, 9), 9) and compute the number of groups as the number of fingers or objects.

C1. Reasoning

A summary of the strategies used by the children to solve the problems is given in Table 4.

Participants used both modelling and counting strategies, and in some cases mental calculation. Moreover, the fact that the children worked in teams induced the appearance of a new strategy, S1*, based on strategy S1 (in which the objects were allocated one-by-one until none was left, with the number of objects in the resulting groups providing the solution). S1* differed from that approach in that, as children worked in teams rather than individually, each team member allocated one object to each group. When the children failed to follow a consistent order, the resulting allocation was unequal and had to be adjusted. Two types of adjustment were used. In one, the children arranged the objects allocated to each group in columns to visualise the uneven distribution, reorganising the objects in keeping with the height of the columns. In the other, as the objects allocated were not arranged linearly, the number allocated to each group had to be counted to make the adjustment (Figure 4).

Another significant finding was that problems of the same type (partitive or quotative) were solved using different strategies. For instance, all the teams used strategy S1 or S1* to solve the (partitive) Pirate1 problem, whereas other strategies were brought into play to solve likewise partitive Ferry2. Other variables, such as the materials used or the context, were believed to prompt the use of one or another strategy. One clear example lies in Loading1, where the subjects interacted with a lorry that carried tiles by road. This quotative problem was solved using partitive strategies, on which the tiles were placed on the lorry and distributed among the various buildings. In contrast, quotative problem Ferry1, involving the carriage of cars across a river by a ship, was solved with quotative strategies: the cars were arranged into groups because the ship's unwieldy size made it difficult to move.

Table 4. Strategies used by six teams of children to solve the problems.

	Team 1	Team 2	Team 3	Team 4	Team 5	Team 6
Pirate1 Partitive	S1*	S1*, S2	S1*	S1*	S1	S1*
Pirate2 Partitive	S1*	S4, S1, calculated mentally	S4	S4	S1*	S4
Loading1 Partitive	S1		S1		S1	S1*
Loading2 Quotative	S1*	SO, S5, calculated mentally		S4, S5, calculated mentally		
Ferry1 Quotative calculated mentally	S5 S7, S9, calculated mentally	S9, S7 S7, calculated mentally	S7, S5			
Ferry2 Partitive		Calculated mentally	Calculated mentally	S7 and S5	S4 and S5	

Note: S1* = strategy based on strategy S1; Blank cells = exercise not done.

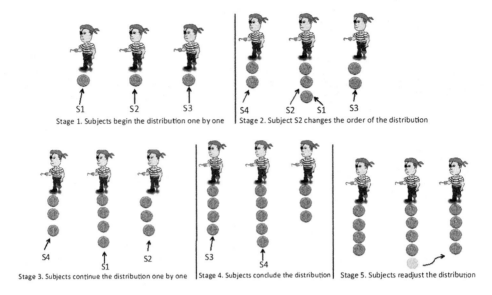

Figure 4. Strategy S1* in problem Pirate1.

In some of the exercises, some teams verbalised their answer, alluding to mental calculations they failed to explain. When solving the Loading2 problem, in addition to using the strategies described, team 2 drew from the similarity with the preceding problem. That indication of some degree of generalisation of the problem structure was illustrated in the following exchange.

Teacher (explaining Loading2): 'you need three tiles for each roof. There are eight buildings. How many won't get tiles? I don't know how many tiles there are. Count them.'

Child 1:	'thirteen.'
Teacher:	'thirteen? Are you sure? Count again.'
Child 1:	'there are twelve.'
Teacher:	'what's going to happen, then?'
Child 1:	'four buildings won't get tiles.'
Child 2:	'four won't get a roof.'
Teacher:	'how did you figure that out?'
Child 2:	'since we did the playmobile thing before and there were twelve coins ... '
Child 1:	'it's the same.'
Child 2:	'and there were four people, and now there are eight, so four buildings won't get a roof.'

The experience consequently showed that the problems used encouraged reasoning and accommodated several solving strategies.

C2. Context

The children were familiar with the elements used, pirates, lorries and boats, thanks to their presence in the media, stories or games. As ferries might be regarded as less common, the exercises were posed around more familiar objects. In the first problem, the context was the distribution of a treasure among pirates; in the second, a lorry delivering building tiles; and in the third, a boat to move cars. Nothing in the children's reactions denoted unfamiliarity with the situations, although they may have been less accustomed to moving cars on a boat.

C3. Challenge

Oral representation was used in all cases when introducing the problems to the children, together with the physical materials or manipulatives. The toys chosen, as shown in Figures 1, 2 and 3, were attractive and familiar to the children. With the exception of the boat in the ferry problem, which had to be large enough to accommodate the cars, all the materials could be readily handled. Some of the subjects opted to move the cars alone, instead of loading them on the boat (Figure 4).

C4. Multiple solutions

Childrenś answers were more than just short replies: they included an explanation of the criterion or procedure followed. The four levels of solutions identified in the replies are discussed below.

In the first level children solved the problem mentally, giving an oral reply without handling the material. An example follows.

Child:	'I'd give them three each.'
Teacher:	'why? How did you come up with that answer?'
Child:	(thinking)
Teacher:	'how did you?'
Child:	'it just occurred to me.'
Teacher:	'how did you share them out?'
Child:	'since there were twelve coins, they each got three.'

In a second level, subjects replied first and then used the material to verify the answer. One of the girls replied to Ferry1 with the following explanation.

Child:	'four times.'
Teacher:	'how did you figure that out?'
	Child (picking up the cars). 'Take three and ship them. Take [another] three and ship them (grouping the cars three-by-three). Take [another] three and ship them and take three [more] and ship them.'

On a third level, the solution was the result of handling the materials. Children used the material to solve the problem and then gave an oral reply.

The fourth level consisted of replies based on the materials only. The solution was displayed using the manipulatives, with no oral reply. The children regarded the problem as solved after handling the respective materials.

The data also showed that children depended on the teacher's evaluation to corroborate that theirs was the right solution.

C5. Expandability

Partitive and quotative division were the underlying mathematical structures in the problems. Although the children had not previously worked with the division, they had mastered the following skills.

- They could compare amounts, count using their fingers or objects, perform simple mental calculations, count two-by-two, three-by-three …
- The were able to add.

Teacher:	'the girl with the pony tail, how many does she have?'
Several children:	'two.'
Teacher:	'and the captain, how many does he have?'
Several children:	'six.'
Teacher:	'is that fair?'
Several children:	'we took two away from him, two and [then another] two.'

- They could subtract.

'And there were four people, and now there are eight, four buildings won't get a roof.'

- They knew multiplication terminology:

| Teacher: | 'three times four then?' |
| Child 1: | '12.' |

The children used the solution to one problem to solve others with different data, denoting an ability to generalise.

Teacher:	'how many tiles do we need then?'
Child 1:	'more.'
Teacher:	'how many more?'
Child 1 (thinking as he gazes upward):	'twelve.'
Teacher:	'for what?'
Child 1:	'so each building can have three.'

One of the teams even generalised the problem structure, identifying a relationship between the problems posed.

Teacher:	'you need three tiles for each roof. There are eight buildings. How many buildings won't get fixed? I don't know how many tiles there are. Count them.'
Teacher:	'some are missing.'
Child:	'four buildings won't get tiles.'
Teacher:	'four won't get a roof.' 'How did you figure that out?'
Child:	'since before we did the playmobile thing and the coins, and there were twelve'. 'And it's the same.' …

C6. Comprehensibility

The language used was familiar to children, who harboured no doubts about problem-wording or the action to be taken to solve them. Moreover, they were able to reach solutions with the materials provided. They verified their solutions by physically handling the materials.

Teacher:	'does everybody understand?'
All:	'yes.'
Teacher:	'what do you need to do?'
Child:	'put tiles on the roof.'

The children showed no signs of 'drawing a blank' when confronted with the problems. Although they had difficulties in verbalising their solving procedures, they were able to outline them. In some cases, different strategies used by children on the same team were even identified as being the same.

Teacher:	'[Child's name] found a solution. Why?'
Child 1:	'I counted. Three cars each time, I say one (counting on his fingers).'
Teacher:	'another three cars, you said.'
Child 1:	'three (signing with his fingers).'
Teacher:	'that seems to be to be a very good way to go about it.'
Teacher:	'how did you go about it?'
Child 2:	'counting the cars. To see if there were twelve. If there are twelve, it takes four times.'
Teacher:	'what did you do?'
Child 2:	'I was going to count three-by-three.'

5. Discussion

Initially, the children's actions corroborated the expectations about the theoretical characteristics attributed to the problems. Two sets of characteristics were identified, depending on the child's mathematical activity and understanding of problem structure.

The first set was associated with diversity in students' mathematical activity, such as posing several solving strategies (C1), different representations (C3) or different levels of solutions (C4). Whilst such diversity might be attributed, a priori, to a given problem, proof that the problems proposed favoured such a wealth of approaches was furnished by the present experience. More than one strategy, representation and solution level were detected in them all. Even strategies not predicted in the literature were observed, as a result of working simultaneously with several children. This set of characteristics supports Matalliotaki's (2012) findings, for at these ages children find it difficult to solve problems without some manner of representation, which is one of the factors that fuels their progress in developing solving strategies.

The other set of characteristics was associated with understanding the problem, its structure and reflection on the solution proposed. The contexts used were familiar (C2) and the language understandable (C6) and both proved to be suitable for introducing the meaning of a new mathematical concept (here, partitive and quotative division) based on prior knowledge (C5). All the teams proposed solutions and were able to solve the same problems with different data. That confirmed earlier research findings on the benefits of a problem-solving approach to mathematics teaching in mathematics learning (Cai 2010).

6. Conclusions

This paper synthesises the characteristics associated with problems suitable for pre-primary education, drawing from prominent earlier research on the subject (Lesh et al. 2013; Nelson and Kirkpatrick 1975; Van de Walle 2003; Yee 2009). The characteristics proposed were applied to analyse three problems put forward by other authors (Davis and Pepper 1992; Nelson and Kirkpatrick 1975) based on the reactions of 26 pre-primary 5–6 year olds. The problems selected were observed to conform to the characteristics defined, confirming their suitability as pre-primary problems. Moreover, the list of characteristics selected proved to be operational for the ex-ante recognition of the features of problems for that stage of schooling, a matter of practical utility, given that problem selection is one of teachers' major tasks (Britz and Richard 1992; Lester and Cai 2016).

The present empirical results showed that problem characteristics depend on two main factors: mathematical activity, which motivates children, and problem structure, which favours understanding the approach to, engagement in and verification of the solutions.

The variety of solving strategies, representations and levels of solutions observed was consistent with other findings on the use of multiplicative structures (Davis and Pepper 1992; De Castro and Hernández 2014; Desforges and Desforges 1980; Matalliotaki 2012).

The three problems studied entailed mathematical ideas, in which the meaning of division was shown to be equitable distribution, establishment of quotas and reiterated subtraction. The classroom experience revealed the presence of characteristic C1, reasoning, although differences were observed among the problems, attributed to the use of materials and working with several children at the same time. The presence of reasoning reported in earlier studies (De Castro and Hernández 2014) was confirmed here by the prevalence of modelling strategies in the first problem, in which all the objects were first counted and then distributed one-by-one until none was left. Conversely, teamwork and children's reactions to the objects to be distributed prompted strategies not identified in earlier research involving individual work (Carpenter et al. 1999; Davis and Pepper 1992). These new approaches favoured readjustment strategies and trial and error cycles, in which the initial allocation in problem 1 of more than the correct number of coins to one pirate was rectified by the necessary rearrangements. A similar situation arose in problem 2, in which the same strategies were observed even in the quotative exercise. Problem 3 was characterised by a prevalence of quotative strategies, however.

The materials provided also had a significant effect on children' use of one strategy or another. In problem 1, linear arrangement of the coins facilitated comparison with no need for counting. In problem 2, the presence of a lorry favoured one-by-one sharing over prior grouping of the elements to be distributed, whereas in problem 3 the physical unwieldiness of the boat prompted deployment of the latter strategy.

The representations used were conditioned by the material provided and the format of the oral reply called for by the teacher. The problems featured characteristic C3 (Challenge), as they engaged children in seeking a solution, which resulted from handling the materials. When the material was less convenient for representing the situation, such as in the ferry problem, oral representation prevailed. Children's prior experience in solving two problems with similar structures may have improved their performance in the third (Matalliotaki 2012). Such a variety of representations favoured the presence of multiple solutions, which were not short answers. Up to four levels of solutions were identified, depending on whether the answer was obtained from mental calculation only, the manipulatives or combinations of the two.

The classroom experience showed that both real and fictitious contexts were familiar to the children, although shipping cars on a ferry was a somewhat less ordinary situation for them. The language used and actions to be performed were understandable and in all cases, the children were able to validate their answers by physically distributing the materials. In some cases, in addition to expressing their confidence in the solution found, they recognised several of the strategies used as equivalent. The experience revealed that these children generalised the mathematical structure involved, and some teams even identified similarities among the problems. Despite having not been taught to divide, their knowledge of counting and arithmetic operations enabled them to deploy strategies apt for solving partitive and quotative division problems. That would support children's ability to understand and solve such problems much earlier than assumed in school curricula (De Castro and Hernández 2014), providing the problems are posed in a manner suited to their age. There may be two explanations for that finding: one may have to do with the practice acquired by the children after solving more than one problem, further to Matalliotaki's (2012) contention that children's performance improves when solving problems with similar structures. The other explanation lies in some children's natural mathematical talent. As the present evidence is insufficient to support either explanation, however, further research is necessary.

Two main contributions are deemed to stem from this study. First, it synthesises the characteristics that define problems apt for pre-primary education, furnishing an enhanced operational approach to analysing new proposals. Second, the empirical findings highlight the role of working with groups of children and of the materials used in favouring strategic diversity and several levels of solutions and representations. Such diversity may be favoured by the problems themselves. Nonetheless, the ideas put forward by Britz and Richard (1992) and Lester and Cai (2016) on the teacher's importance in the emergence of different strategies are supported by the present findings, which revealed the significance of variables such as presenting the children with materials, the size of the objects used and working in teams. Quotative problems were solved with partitive strategies because of the format of the materials, revealing their decisive role in the meanings of the mathematical concepts introduced.

This study is subject to limitations, in particular as regards the small number of children involved and the use of three very specific problems. Nonetheless, the experience described is deemed to be repeatable in other classrooms for comparison with the results reported here. Future lines of research would include analysing the validity of other problems in terms of the characteristics proposed and contrasting the empirical findings, particularly in connection with the impact of materials on solving strategies.

Disclosure statement

No potential conflict of interest was reported by the authors.

References

Agre, G. P. 1982. "The Concept of Problem." *Educational Studies* 13 (2): 121–142.

Britz, J., and N. Richard. 1992. *Problem Solving in the Early Childhood Classroom.* Washington, DC: NEA.

Cai, J. 2010. "Helping Elementary School Students Become Successful Mathematical Problem Solvers." In *Teaching and Learning Mathematics. Translating Research for Elementary School Teachers,* edited by D. V. Lambdin, and F. K. Lester, 9–14. Charlotte, NC: NCTM.

Carpenter, T. P., E. Fennema, M. L. Franke, L. Levi, and S. B. Empson. 1999. *Children's Mathematics. Cognitively Guided Instruction.* Portsmouth, NH: Heinemann.

Castro, E., and E. Castro. 2016. *Enseñanza y aprendizaje de las matemáticas en Educación Infantil* [Teaching and Learning Mathematics in Early Childhood Education]. Madrid: Pirámide.

Charlesworth, R., and S. A. Leali. 2012. "Using Problem Solving to Assess Young Children's Mathematics Knowledge." *Early Childhood Education Journal* 39 (6): 373–382.

Clements, D. H., and J. Sarama. 2007. "Early Childhood Mathematics Learning." In *Second Handbook of Research on Mathematics Teaching and Learning*, edited by F. K. Lester, Vol. 1, 461–556. Charlotte, NC: NCTM.

Clements, D. H., and J. Sarama. 2013. "Solving Problems: Mathematics for Young Children." In *Handbook of Research-Based Practice in Early Education*, edited by D. R. Reutzel, 348–363. New York, NY: The Guilford Press.

Davis, G., and K. Pepper. 1992. "Mathematical Problem Solving by pre-School Children." *Educational Studies in Mathematics* 23 (4): 397–415.

De Castro, C., and E. Hernández. 2014. "Problemas verbales de descomposición multiplicativa de cantidades en educación infantil." [Verbal Problems of Multiplicative Decomposition of Quantities in Kindergarten]. *PNA* 8 (3): 99–114.

Desforges, A., and C. Desforges. 1980. "Number-Based Strategies of Sharing in Young Children." *Educational Studies* 6 (2): 97–109.

Kilpatrick, J. 1980. "What is a Problem?" *Problem Solving* 4 (2): 1–5.

Lesh, R., L. English, C. Riggs, and S. Sevis. 2013. "Problem Solving in the Primary School (K-2)." *The Mathematics Enthusiast* 10 (1&2): 35–60.

Lesh, R., and J. Zawojewski. 2007. "Problem Solving and Modeling." In *Second Handbook of Research on Mathematics Teaching and Learning*, edited by F. K. Lester, Jr. Vol. 2, 763–804. Charlotte, NC: NCTM.

Lester, F. K., and J. Cai. 2016. "Can Mathematical Problem Solving be Taught? Preliminary Answers From 30 Years of Research." In *Posing and Solving Mathematical Problems*, edited by P. Felmer, E. Pehkonen, and y J. Kilpatrick, 117–135. New York, NY: Springer.

Matalliotaki, E. 2012. "Resolution of Division Problems by Young Children: What are Children Capable of and Under Which Conditions?" *European Early Childhood Education Research Journal* 20 (2): 283–299.

Mayer, R. E., and M. C. Wittrock. 2006. "Problem Solving." In *Handbook of Educational Psychology*, edited by P. A. Alexander, and P. H. Winne, 287–303. New York, NY: Routledge.

McMillan, J. H., and S. Schumacher. 2010. *Research in Education: Evidence-Based Inquiry*. 7th ed. Boston, MA: Pearson.

MEC. 2006. *Real Decreto 1630/2006, de 29 de diciembre, por el que se establecen las enseñanzas mínimas del segundo ciclo de Educación Infantil* (BOE, n° 4, pp. 474–482). Madrid: Ministerio de Educación y Ciencia.

Mulligan, J. T., and G. Vergnaud. 2006. "Research on Children's Early Mathematical Development. Towards Integrated Perspectives." In *Handbook of Research on the Psychology of Mathematics Education*, edited by A. Gutierrez, and P. Boero, 117–146. Rotherdham: Sense Publishers.

NAEYC, & NCTM. 2010. *Early Childhood Mathematics: Promoting Good Beginnings. A Joint Position Statement*. NAEYC. http://www.naeyc.org/files/naeyc/file/positions/psmath.pdf.

NCTM. 2000. *Principles and Standards for School Mathematics*. Reston, VA: NCTM.

Nelson, I. D., and J. Kirkpatrick. 1975. "Problem Solving." In *Mathematics Learning in Early Childhood. Thirty-Seventh Yearbook*, edited by J. N. Payne, 69–94. Reston, VA: NCTM.

OECD. 2016. "What are the Benefits From Early Childhood Education?" *Education Indicators in Focus* 42: 1–4.

Pólya, G. 1945. *How to Solve it*. New Jersey, NY: Princeton University Press.

Reys, R., M. M. Lindquist, D. V. Lambdin, and N. L. Smith. 2009. *Helping Children Learn Mathematics* (9th ed.). Danvers, MA: Wiley & Son.

Schoenfeld, A. H. 1985. *Mathematical Problem Solving*. New York, NY: Academic Press.

Schoenfeld, A., and D. Stipek. 2011. *Math Matters: Children's Mathematical Journeys Start Early*. Conference Report, Berkeley, CA. http://www.earlymath.org.

Van de Walle, J. A. 2003. "Designing and Selecting Problem-Based Task." In *Teaching Mathematics Through Problem Solving: Prekindergarten-Grade 6*, edited by F. K. Lester, and R. I. Charles, 67–80. Reston, VA: NCTM.

Yee, F. P. 2009. "Mathematics Problem Solving." In *Teaching Primary School Mathematics: A Resource Book*, edited by L. P. Yee, and L. N. Hoe, 65–94. Singapore: McGraw-Hill.

9 Learning through play – pedagogy and learning outcomes in early childhood mathematics

Franziska Vogt, Bernhard Hauser, Rita Stebler, Karin Rechsteiner and Christa Urech

ABSTRACT

Whilst research underlines the importance of early mathematics in kindergarten, practitioners need effective and innovative approaches to pedagogy. Currently, very different approaches are deployed from an instructional, educator-led approach based on training programmes to a play-based approach. This intervention study examines the effects on the mathematical competency of these two pedagogies. Thirty-five kindergarten educators and 324 six-year-old children were randomly assigned to either a training programme, a play-based approach with card and board games or to the control group. Educators' views on the interventions were gathered in semi-structured interviews. The results indicate higher learning gains overall for the play-based approach. Differentiated effects were found as tendencies: children with low competencies tend to gain more from training programmes compared to no intervention; children with high competencies gain more from the play-based approach than the training. Educators evaluated the play-based intervention with card and board games as better suited to children's diverse needs.

Introduction

Early mathematical competencies are highly relevant for later education outcomes (Duncan et al. 2007; Grüssing and Peter-Koop 2008). Whilst there is a growing awareness that children need to already be supported in their mathematical learning in kindergarten, there is little consensus about the best pedagogical approach. Kindergarten educators may emphasise that mathematical activities need to be embedded in everyday situations (Gross and Rossbach 2011) or that early learning needs to be based on play, even though the understanding of play itself varies (Gasteiger 2015). Furthermore, kindergarten educators might use a training programme for mathematics, to ensure that mathematical competencies are explicitly fostered. Little research exists on the effectiveness of these approaches, as they have not yet been systematically compared regarding the learning gains for all children and for outcomes of children with differing levels of competencies. The research project presented here compares the effectiveness of a play-based approach with card and board games (Hauser et al. 2015) with a training programme (Krajewski, Nieding, and Schneider 2007) with a control group. In addition, educators' acceptance of an approach is important for effective implementation. Therefore, educators' views need also to be taken into account. This paper addresses the following research questions: how does the play-based approach with card and board games compare with the training programme regarding children's mathematical learning gains? Are there differentiated effects for children with differing mathematical competencies? What are educators' experiences with and views on the play-based approach and the training programme?

Theoretical framework: mathematical competencies and approaches for kindergarten

The literature review starts with research findings on the relevance of mathematics education in kindergarten, followed by a focus on certain aspects of mathematical competencies. Thereafter, consideration is given to research on early childhood educators' attitudes towards mathematics. Then, several approaches to early mathematics are outlined, followed by the discussion of the innovative potential of play in the teaching of early mathematics.

Mathematical competencies matter

Mathematical competencies at kindergarten are highly relevant for learning outcomes at school. In a meta-analytic regression, maths competencies in kindergarten, i.e. number recognition, number sequence, counting, ordinality, relative size, addition and subtraction were found to be the strongest predictors for later school achievement, reaching an average effect size of .34 compared with early reading (.17), attention skills (.10) and socio-emotional behaviours (no effect) (Duncan et al. 2007). Children with low mathematical competencies in kindergarten are most likely to experience difficulties with maths at school (Dornheim 2008). Quantity–number–competencies predict later mathematical competencies beyond number words whilst phonological awareness does not (Krajewski and Schneider 2009). Children's mathematical competences differ considerably in kindergarten, which is also due to differences in the home learning environment (Anders et al. 2012; Sonnenschein and Galindo 2015). In order to enhance opportunities for all children, regardless of their family background, kindergarten needs to foster mathematics intentionally (Grüssing and Peter-Koop 2008) and children need to be provided with learning opportunities which meet their diverse educational needs (Gasteiger 2015). As with other subject areas, the quality of teaching is crucial but also highly variable (McCray and Chen 2012). Engel et al. (2016) linked the time spent on maths, as reported by the educators, with children's maths achievement and found no correlation. They concluded that educators focus on curricular content, which is not sufficiently challenging for most children, e.g. counting and shapes.

Competencies that need to be fostered

Mathematical learning opportunities are needed which are challenging, appropriate and adaptive to the heterogeneous needs of young children. Three-to-four-year-old children need guidance to notice quantitative relationships in daily life and to begin representing 'information about pattern, shapes, space and number' (McCray and Chen 2012, 292). As Sarama and Clements (2009) highlight, the aim is to foster 'overarching' mathematical competencies, which are core for mathematics and in line with children's thinking. Amongst these, the quantity–number–competencies are highly relevant as longitudinal studies found: quantity–number–competencies at school entry are the strongest predictor for mathematical achievement in third grade (Krajewski and Schneider 2009) and pre-school quantity–number–competencies explain up to 41% of the variance in mathematical competencies at primary school, whereas general cognitive ability only up to 10% (Dorn-heim 2008). The theoretical model of the development of quantity–number–competencies (Krajewski 2003) provides an orientation on the development of these mathematical competencies from the basic level 'number word sequence isolated from quantities' to 'quantity to number word linkage' to the 'concept of number relationships' (Krajewski and Schneider 2009, 517ff.).

Early childhood educators' beliefs on mathematical learning

Educators' beliefs are likely to influence the teaching of mathematics in kindergarten. Their feeling of self-efficacy regarding mathematics is related to the importance they assign to the subject in kindergarten (Brown 2005). Educators might be worried that maths is 'not fun' (Lee and Ginsburg 2009) and express negative feelings towards mathematics (Benz 2012), possibly shaped by their own, often negative, school experience (Anders and Rossbach 2015). Other findings indicate a positive attitude towards maths amongst early years' educators (Chen et al. 2014; Thiel 2010). Link, Vogt, and Hauser (2017) found in a comparison of educators' beliefs regarding fostering mathematics in kindergarten between Austria, Germany and Switzerland that the Swiss educators agree more strongly to an intentional approach to mathematics in kindergarten than the German and Austrian educators.

Different approaches for early maths

Taken that challenging, appropriate and adaptive mathematical learning opportunities are required for kindergarten and given that quantity–number–competencies are particularly relevant for later learning, educators need to decide on the best approaches to support the acquisition of these competencies in kindergarten. Schuler (2008) lays out several decisions, which educators face, amongst them the decision between (i) an instructional programme versus free learning arrangement, (ii) specific fostering of children at risk versus fostering for all children and (iii) focussing on domain-specific competencies only versus a broader approach. Traditionally, mathematics was not at the centre of curricular attention in kindergarten and educators emphasised a situated approach whereby mathematical competences are applied to everyday situations, i.e. counting the children present, comparing quantities when sharing fairly, performing one-to-one correspondence when laying the table. However, the emphasis on early learning leads to a shift in kindergarten practice. A range of learning materials are available, many of them requiring a specific time frame for focused mathematical activity, many of them designed as a training programme, with a set order of units and exercises focussing on defined skills, designed to be worked through in the given order, delivered in an educator-led instructional group setting. Educators are often not in favour of using training programmes, as they are seen as too much like school (Jörns et al. 2014). The rise in instructional, school-like training programmes for kindergarten raises the question, whether a highly educator-centred, instruction-focused approach is best suited for children of kindergarten age or whether a play-based approach would be more appropriate (Hauser 2005).

Play as an innovative approach

Innovative approaches to early mathematics should not only be developmentally adequate and effective, but also compatible with the kindergarten pedagogy. As kindergarten children are highly motivated to learn, but not in a formal, instructional way, play can be regarded as a powerful vehicle for learning (Hauser 2005). Play can be defined as activities that 'are fun, voluntary, flexible, involve active engagement, have no extrinsic goals, involve active engagement of the child, and often have an element of make-believe' (Weisberg, Hirsh-Pasek, and Golinkoff 2013, 105).

Play and playfulness are at the core of early childhood education (Singer 2013), although educators are not always aware of their role in fostering play (Bodrova 2008; Vu, Han, and Buell 2015). It is important to distinguish between activities, which are play-based and adult-initiated activities, which resemble school-like tasks (Bergen 2015). Weisberg et al. (2015, 10) coined educator-led educational activities disguised as play as 'chocolate-covered broccoli'. Wood (2009) highlights the need to distinguish between the different forms on the continuum between free play and no play. Such a clarification is sought with the concept of 'guided play': 'guided play sits between free play and direct instruction' (Weisberg, Hirsh-Pasek, and Golinkoff 2013, 105) and consists of adults structuring of the play environment but leaving control to the children within the environment (Weisberg et al. 2015).

Innovative approaches to early mathematics could draw on play, be it role-play (van Oers 2010) or board and card games. It needs to be recognised that role-play, or pretend play, requires much time for the children to set up the play-frame, to engage with the play and develop it (Bergen 2015). As for board and card games, several studies found them to be effective in the acquisition of mathematical competencies (Gasteiger 2015; Jörns et al. 2014; Kamii and Kato 2005; Ramani and Siegler 2008; Schuler 2013). Gasteiger, Obersteiner, and Reiss (2015, 232) highlight that not only the concept of 'play' is deployed differently, but also 'games'; consequently, they propose a 'continuum from games designed for the purpose of entertainment only to targeted instruction with only few entertaining features'. Four aspects are essential to play-based approaches to mathematics in early childhood education: (i) the 'mathematical content needs to be part of the mechanics of the game'; (ii) needs to be 'correctly presented'; (iii) 'essential for further learning' and (iv) the game needs to be 'appropriate for the individual learning needs of the child' (Gasteiger, Obersteiner, and Reiss 2015, 233f).

Although play is widely acknowledged as an important learning path in early childhood education, little is known about the effectiveness of play in comparison to other ways of learning in early childhood education settings. The project presented here compared a play-based approach with card and board games with a training programme. This paper focusses on the analysis of the interviews and the learning outcomes, as the following main research question is addressed: how does the play-based approach to early mathematics with card and board games compare to the training programme regarding children's learning outcomes and educators' views and pedagogical preferences?

Methods

Research design

The research project compared two intervention groups, a play-based intervention and a training programme, alongside a control group in a pre-post-test quasi-experimental design based on measures of children's mathematical competencies. The two interventions needed to be as comparable as possible regarding content and intervention time.

A training programme – known as being effective from previous testing – *Mengen zählen Zahlen* [Quantity, counting, numbers] was selected (Krajewski, Nieding, and Schneider 2007, 2008). It consists of 24 units of half an hour each, focussing on quantity–number–competencies. It is educator-led, addressing a small group of children, using specific tasks, maths talk and materials.[1] A play-based approach using card and board games was developed by the research team matching the curricular content of the training programme, thus it involves comparing quantities, counting, number recognition and part-and-whole relationship. Some of the games were already available, such as *Halli Galli*; others required an adaptation of the rules or materials, such as *Shut the Box* and *Lining up the Fives*; and some games were developed from scratch for the project, such as *More is More*[2] (Vogt and Rechsteiner 2015). All games were piloted with kindergarten children and evaluated for their suitability in collaboration with three kindergarten educators. For the play-based intervention, the educators were provided with a box of the 10 specific card and board games. The play-based intervention was of the same duration as the training programme: All the children played the card and board games during 24 half-hour units in small groups. In general, the children could choose their co-players and a game, but they were required to play the 'maths games' provided in the box during the half-hour units of the intervention. The educators introduced the games and supported the children. The kindergarten educators of the control group were asked to carry on their practice as always. Widely used ways of fostering mathematical competencies include counting in day-to-day situations, playing with dice and role-play. All kindergarten teachers are required to teach the competencies defined in the curriculum but are free in their choice of pedagogical approach. Regarding research ethics, it can be stated that the research project experimentally varying pedagogical approaches with interventions of eight weeks within the framework of the kindergarten curriculum did not subject children to any disadvantage. Parents and children were informed about the research project and gave full consent to participation.

The educators of both intervention groups received the same general introduction into the learning of mathematics in kindergarten (1 day) and an introduction into either the play-based approach or the training programme (1 day) and two separate follow ups (2 half-day meetings). Data on educator's views were collected at the end of the intervention by an independent researcher using semi-structured problem-centred telephone interviews.

Sample

From the list of all kindergartens in the Canton of St. Gall in Switzerland, kindergarten educators were contacted at random and invited to participate in the research and randomly assigned to one of the groups. In each of the participating kindergarten, the group of children in the last year before entering primary school, i.e. five-to-six-year-old children, participated. The sample included 12 kindergarten educators and 111 children in the intervention group using the training programme *Mengen zählen Zahlen*, 11 kindergarten educators[3] and 91 children in the intervention group implementing the play-based approach and 12 kindergarten educators and 127 children as the control group. The children's mean age was 6 years and 3 months, with no differences between the three goups ($F[2,325] = 1.400$; $p > .05$).

Instruments and data analysis

The quantitative research instruments and the statistical procedures are outlined first, followed by the qualitative instruments and the qualitative data analysis.

Quantitative instruments: the mathematical competencies were measured using the test *Zahlenstark*, developed by Moser and Berweger (2007) and employed for a large-scale evaluation in Switzerland (Moser and Bayer 2010). The test is compatible with the Krajewski model of the development of mathematical competency, integrating approaches of measuring mathematical competencies used by Krajewski (2003), Van den Heuvel-Panhuizen (1995) and Moser Opitz (2001). The test involves tasks on ordinality, cardinality, quantity, number knowledge and first arithmetic operations, often proceeding from tasks illustrated with images and embedded in an everyday story to numbers-only representations. Research assistants conducted the test one-to-one with each child on the premises. Cognitive abilities were measured using two subtests from CFT1 (Weiss, Cattell, and Osterland 1997). Parents completed a questionnaire with questions on the socio-economic background of the family, languages spoken at home and the home learning environment.

Quantitative data analysis: In order to ascertain whether the groups were comparable, an analysis of variance (ANOVA) on children's age, cognitive ability, socio-economic status of the family, migration background and pretest mathematical competencies was conducted. Then the results of the mathematical competencies tests were compared by applying an analysis of variance with repeated measures.[4] Differentiated outcomes were explored dividing the children into three groups, a third of the overall sample according to pretest mathematical competency forming a high-level group, medium-level group and low-level group, and ANOVAs were performed on each subgroup.

Qualitative instruments: The kindergarten educators of both intervention groups were interviewed by an independent researcher after the intervention. The interviews were held on the phone and had a duration of 30–40 minutes. The educators were first asked to imagine a scenario, whereby they would explain to a colleague, what the project they participated in was about. Then they were asked to describe how they implemented the intervention, how the children engaged with the intervention and – as an overall concluding evaluation – whether they are likely to use the play-based approach or the training programme in the future. The semi-structured problem-centred interviews were audiotaped and transcribed in full.

Qualitative data analysis: The transcripts were first analysed using qualitative content analysis with the software MAXQDA (Kuckartz 2010). Whilst as a first step, the interviews were carefully analysed in order to identify problems with specific games, the re-analysis of the interviews presented in this paper focusses on educators' views and experiences. Furthermore, the response to the narrative scenario question at the beginning of each interview as well as their sense-making on pedagogical approaches to mathematics in kindergarten was analysed in detail for each educator focussing on discourse (Kruse 2015).

Results

The results to the question addressed in this paper – how the play-based approach to early mathematics compares with the training programme regarding educators' views and pedagogical preferences and children's learning outcomes – will be set out as follows: first, the quantitative results on children's mathematical competencies and, then the qualitative results on educators' views.

Results regarding children's learning gains

The three groups proved to be comparable at pretest, as no differences were found regarding cognitive abilities, socio-economic status and migration background (details see Hauser et al. 2014). The mathematical competencies of the children at pretest do not differ between the three groups (Table 1). The treatments have an influence on mathematical learning gains, the ANOVA for repeated measure reveals a significant interaction between time and group ($F[2,321] = 4.04$; $p = .019$; partial $\eta^2 = .025$) (Table 1) with higher learning gain for the play-based intervention compared with the control group (Bonferroni post hoc $p = .015$). The calculation of the effect size results in Cohen $d = 0.30$, a small size effect (Cohen 1988).

In order to determine whether children might benefit in different ways from the intervention depending on their competency level, the children were divided into three groups according to their pretest maths competency. A third of the group was assigned to the groups of low competency level, middle level and high level, respectively. Table 2 provides the overview of competency measures for the three groups.

The learning gains of the groups according to level differ (ANOVA, $F[2,321] = 11.416$; $p = .000$) and post hoc Bonferroni tests show significantly higher gains for the low-level group compared to both, the middle group ($p = .005$) and the high group ($p = .000$), whereas learning gains of the middle and high group do not differ. For all three level groups, an ANOVA of repeated measure was conducted to detect possible differences

Table 1. Mathematical competencies at pre- and post-test.

	M (SD)[a] pretest	M (SD)[b] post-test	M (SD) learning gains	ANOVA time × group	*Post hoc* Bonferroni
Play-based	63.89 (17.14)	75.24 (17.59)	11.35 (9.15)	$p = .019$;	Play > control,
Training programme	65.21 (19.78)	74.25 (17.89)	9.05 (8.70)		$p = .015$
Control	60.58 (17.89)	68.59 (18.03)	8.02 (7.87)		

[a]No difference between the groups at pretest (ANOVA: $F[2,323] = 1.988$, n.s.).
[b]ANOVA for repeated measure: $F[2;321] = 4.04$; $p = .019$; partial $\eta^2 = .025$.

Table 2. Mathematical competencies of low level, middle level and high level.

Competency level at pretest	N	Mean pretest (SD)	Min pretest	Max pretest	Mean post-test (SD)	Learning gains (SD)
Low	109	43.08 (9.09)	14	54	55.33 (11.78)	12.25 (8.20)
Middle	102	62.81 (4.53)	55	70	71.44 (8.45)	8. 63 (7.85)
High	113	82.55 (10.40)	71	109	89.56 (12.68)	7.01 (8.87)

according to the treatment. In the low level group, a tendency was found ($F[2,106] = 2.84$, $p = .063$, partial $\eta^2 = .051$) with the training programme possibly enabling a higher learning gain than the control group (Bonferroni post hoc, marginal significance, $p = .078$). For the middle group, there is no significant interaction regarding time and group. For the high-level group, a tendency was found ($F[2,110] = 2.77$; $p = .067$; partial $\eta^2 = .048$) with a marginally significant difference between the play-based approach showing higher learning gains than the training programme ($p = .065$) (Table 3).

Results regarding the educators' views and integration into their pedagogy

The results from the qualitative interviews are reported as follows: first, the overall characterisation of the interventions by the educators is analysed; second, educators' assessment on suitability and learning gains is described and third, their experiences regarding the integration of the intervention into their pedagogy is summarised.

Characterisation of the interventions: When asked how they would describe the project to a colleague, the educators of the group with the play-based approach mentioned the keywords 'games' and 'mathematics' and the organisation of the project:

> I am taking part in a project on early mathematical competencies and I find it exciting. We received a whole box with materials and introduced the games and allow the children to work independently afterwards. Several competencies are fostered and it is very varied and exciting and the children are interested. (play7)

The educators of the group with the training programme often used the term 'training programme' (7/12) and emphasised the mathematical content (8/12):

> So I would tell her [the imagined colleague] that I follow a training programme, which I use three times a week with the children … It focusses on the numbers zero to ten, in order to learn the basics, such as what does a number actually mean, or a quantity, and so to build a foundation for later arithmetic. (train11)

Educators' assessment on suitability and learning gains: Almost all educators in both groups mentioned the quality of the material provided. The games as well as the training programme contain material made of wood, which for some of the educators is important.

Table 3. Learning gains of low level, middle level and high level according to treatment groups.

	M (SD) learning gains play	M (SD) learning gains training	M (SD) learning gains control	p time × group	Post hoc Bonferroni
Low	13.16 (8.73) n = 32	14.21 (6.99) n = 34	10.23 (8.34) n = 43	.063	Training > control p = .078
Middle	11.09 (9.11) n = 23	8.88 (8.50) n = 32	7.26 (6.48) n = 47	n.s.	
High	9.82 (9.53) n = 34	5.18 (8.13) n = 44	6.57 (8.66) n = 35	.067	Play > training p = .065

All educators would use the material again in the following year. Whereas almost all educators (10/11) plan to implement the play-based intervention in the following year in a similar way, only half of the educators (5/12) would implement the training programme again. Several educators emphasise that they would use the material of the training programme, but adapt the pedagogical approach and only target children with low mathematical competencies:

> I would use it again but would proceed selectively. I would gather the really weak children and work more intensively with them. I would pick out some of the attractive things, for example build the number road, as an offer within free play for all children. (train3)

For the play-based approach, the educators would do the same intervention again. As this educator emphasised, all children benefitted from the intervention:

> I will certainly do it again next year. I thought I would do it exactly the same way. … The weaker children have benefited a lot, but also the strong ones. (play5)

For the training programme, all educators criticised that it did not meet the needs of all children and mentioned the problem of boredom; half of the educators said, repeatedly, that the children became very bored:

> They really did not find it very cool anymore, they did not engage anymore because they were not sufficiently challenged. (train9)

Several educators also expressed their concern that the children had to sit and listen for a long time. For the play-based intervention, boredom was not mentioned at all, only one educator felt, that the children's motivation stayed not as high as she hoped.

Integration of the intervention into their pedagogy: Several educators from the play-based approach expressed that the pedagogy of kindergarten should not entail as many programmes. While the training programme on phonological awareness is widespread and sometimes deemed compulsory, the educators would not want more training programmes in addition to the one they already have to implement. Some educators assigned to the training programme expressed that they would have wished to be in the play-based group as this would suit their pedagogy better. They expressed concern that the children had no choice of the activity in the training programme and that it was too school like:

> I think now that for me, it would have been more important to be in the other group, as there was more freedom and fun things, where the children were able to choose themselves: what is cool? What am I able to do mathematically? … This would be more what kindergarten is about. … . Fun got lost a bit, as they had to do what they maybe did not want to do, and what they will have to do a lot in school later. (train9)

A few educators expressed that they were first concerned, that mathematics should not really be a topic in kindergarten, but then discovered that the children liked the games:

> These are really games which the children like a lot. … it fosters several areas of competencies without the children noticing, as it is all playful … I just think that mathematics should not be as present in kindergarten. (play3)

Both interventions were integrated into classroom routines, so, for example, the way of allocating games:

> I have a photograph of each child … and I assigned them to the games. So their photo was next to a place and they came into the room and looked around, "where is my photo", and then right away began to play (play1).

For both approaches, educators appreciated that there is no theme or fantasy world suggested:

> I think the material [of the training programme] is beautiful and very well suited. … it is not so over-done with knick-knacks or gimmicks, but it is very straight forward. (train7)

Discussion

The comparison of pre- and post-test, with eight weeks of intervention in between, showed a significantly higher learning outcome for the group of play-based mathematics compared to the traditional kindergarten, but no effects for the training programme. This contrasts with other evaluations where this training programme resulted in significant learning outcomes (Krajewski, Nieding, and Schneider 2008). The significant learning gains for the play-based approach as compared to the control group underlines that it is possible to obtain learning gains with an approach of guided play, using card and board games. So far, research into play-based approaches has compared effects with a control group, but not with another treatment, and found games to be effective (Gasteiger 2015; Jörns et al. 2014; Kamii and Kato 2005; Ramani and Siegler 2008). As there are no significant differences between the play-based intervention and the training programme, it can be stated that this play-based approach is at least as effective as a highly educator-led, instructional training programme. The play-based approach adheres to the idea of 'guided play' (Weisberg, Hirsh-Pasek, and Golinkoff 2013): free choice for children and an emphasis on peer learning, the play is guided as the card and board games are geared to specific mathematical competencies within an educator-structured learning arrangement and controlled time frame.

Whilst the training programme delivered marginally significant learning gains amongst children with low levels of competencies – compared to the control group – the training programme is not adapted to children with higher levels. It can be concluded that training programmes delivered to the whole group run the risk of having a detrimental effect, as the majority of the children are taught curricular content they already know (Engel et al. 2016) and so become bored.

The educators' assessment of learning gains based on their day-to-day observation corresponds with the differentiated quantitative results. They also described that the training programme was mainly beneficial for children with very low competency, but that the play-based approach served all children, from low to high competency. The play-based approach was evaluated more positively, as it was considered more fun and less school like. It was, therefore, more compatible with educators' pedagogical beliefs, similar to findings of views on the emphasis on fun (Lee and Ginsburg 2009), positive emotions (Anders and Rossbach 2015) and 'true' and 'entertaining' play (Gasteiger 2015). For both interventions, educators appreciated that the interventions did not require a specific topic or fantasy world (Schuler 2013) but that they could integrate the approach into their day-to-day routines.

Conclusion

The study effectively demonstrates that innovative approaches to early maths can be successfully based on play, particularly on card and board games, specifically developed with regard to quantity–number–competencies as defined in the curriculum. The specific potential of card and board games can be found in the opportunity of performing mathematical activities over and over again and the motivation of a peer-group setting, whereby co-players monitor and support each other (Stebler et al. 2013). The play-based approach developed here is not free play but guided play (Weisberg, Hirsh-Pasek, and Golinkoff 2013). The aim of this study was to compare the effectiveness of two different pedagogies in kindergarten aimed at the identical quantity–number–competencies. Further research might seek to establish play-based approaches to mathematics in kindergarten which target a wider range of mathematical competencies.

The results of this study highlight the importance of meeting individual children's diverse needs (Garrote, Moser Opitz, and Ratz 2015; Gasteiger 2015). Children's mathematical competencies in kindergarten in the year before starting school are very diverse (Dornheim 2008; Krajewski and Schneider 2009). More instructional training programmes with a 'one size fits all' approach fail to challenge and empower every child. However, the findings indicate that a targeted training programme for children at risk is effective together with a range of games for different competency levels for all children's mathematical learning.

As good as learning materials can be – their educational potential can only be realised through good teaching and learning support. Also within the play-based approach, educators perform diagnosis, structure the learning arrangement, provide an impulse, a question and demonstrate strategies for solving the mathematical problems and foster discussions on mathematics amongst the children (Wullschleger and Stebler 2016). In addition, educators' content knowledge can be influential for play-based learning support (Oppermann, Anders, and Hachfeld 2016). In the interviews, the educators expressed their pedagogical beliefs but the possible influence of such beliefs on the learning outcome could not be statistically assessed, due to sample size. Future studies might focus on the play-based approach with a much larger sample to examine educators' competencies and beliefs.

Bringing the results together with the interviews, it becomes apparent, that all educators had a positive attitude towards the interventions. As the interventions progressed, some expressed concern about children becoming bored with the instructional setting of the training programme. The interviews clearly reveal that the educators were more enthusiastic about a play-based approach. Their positive attitude might have been a contributing factor to the learning success of the children found in this study. As there are strong traditions within the pedagogy of kindergarten, it remains important that innovative approaches to mathematics in early childhood not only prove to be effective in terms of learning gains but also that these innovations are easily integrated into the pedagogy of kindergarten. Based on the views expressed in the interviews, as well as their description of their day-to-day experience in kindergarten, a play-based approach to early mathematics certainly has great potential to become an innovation, which will be adopted readily and widely by the practitioners in the field.

Notes

1. The unit 2.5 of the training programme 'Mengen, Zahlen, zählen' (Krajewski, Nieding, and Schneider 2007, 47f) is used to illustrate the training programme. The unit uses the so-called stairs of numbers – a representation of the numbers 1–10 on wooden blocks – structured according to height, on which the number as well as the corresponding amount of points are printed. The educator places the stair of numbers in the middle of the circle of pupils and explains: '7 is less than 8 – pointing to the blocks of the stairs of numbers – less things belong to 7 than to 8'. Then one child after the other is asked a question like: 'what is less, 5 or 4? Why?' Afterwards, each child is asked to take two different blocks, to place them in front of him/her and to say '[number] is more/less than [number]'. The goal of the unit is to recognise the structure of numbers, that the higher number 'is more', contains one more, or several more things than the lower number.
2. As one example of the card and board games 'More is More'(Vogt and Rechsteiner 2015) is described: Each card depicts three structured quantities of points in different colours, with structured representation. Altogether, the 45 cards in the game include quantities in eight different colours. The cards are distributed evenly among players. Each child places their cards as a stack face down in front of them. In the middle of the table lies one card. Children simultaneously and continuously lay open the top card and compare quantities and colours with the central card. If their card shows 'more' of one of the colours than the central card, they lay their card in the middle on top, thus the central card changes continuously. If their card does not depict a quantity which is more and of the same colour than the middle card, the child places that card in a new pile and uncovers the next card. When the first stack is worked through, the child takes the new pile of his/her cards and continues. The first child to discard all his/her cards is the winner. This game aims to develop competency in sub-itising as children are required to compare quantities very quickly without counting in order to maintain speed. The educators are advised that for this game, children at similar competency level should play together. Unlike 'More is More', most other card and board games in the intervention can also be played in heterogeneous groups.
3. Initially, 12 kindergartens were recruited for the play-based approach too, but one educator had to drop out on health grounds.
4. First an ANCOVA was run on the post-test result as the depending variable, including pre-test results and cognitive abilities as covariate, and group as a factor. As cognitive ability did not prove to be a significant covariate, a repeated measure ANOVA was selected as more appropriate (Field 2009).

Acknowledgements

The authors thank the kindergarten educators, as well as the children and their parents, for taking part in this research project.

Disclosure statement

No potential conflict of interest was reported by the authors.

Funding

This research project was supported by the Swiss National Science Foundation [grant number 100014_124485].

ORCID

Franziska Vogt ⓘ http://orcid.org/0000-0002-2023-0431

References

Anders, Y., and H.-G. Rossbach. 2015. "Preschool Teachers' Sensitivity to Mathematics in Children's Play: The Influence of Math-Related School Experiences, Emotional Attitudes, and Pedagogical Beliefs." *Journal of Research in Childhood Education* 29 (3): 305–322. doi:10.1080/02568543.2015.1040564.

Anders, Y., H. G. Rossbach, S. Weinert, S. Ebert, S. Kuger, S. Lehrl, and J. von Maurice. 2012. "Home and Preschool Learning Environments and their Relations to the Development of Early Numeracy Skills." *Early Childhood Research Quarterly* 27 (2): 231–244.

Benz, C. 2012. "Attitudes of Kindergarten Educators about Math." *Journal für Mathematik-Didaktik* 33 (2): 203–232.

Bergen, D. 2015. "Psychological Approaches to the Study of Play." *American Journal of Play* 7 (2): 51–69.

Bodrova, E. 2008. "Make-Believe Play Versus Academic Skills: A Vygotskian Approach to Today's Dilemma of Early Childhood Education." *European Early Childhood Education Research Journal* 16 (3): 357–369.

Brown, E. T. 2005. "The Influence of Teachers' Efficacy and Beliefs Regarding Mathematics Instruction in the Early Childhood Classroom." *Journal of Early Childhood Teacher Education* 26 (3): 239–257.

Chen, J. Q., J. McCray, M. Adams, and C. Leow. 2014. "A Survey Study of Early Childhood Teachers' Beliefs and Confidence about Teaching Early Math." *Early Childhood Education Journal* 42 (6): 367–377. doi:10.1007/s10643-013-0619-0.

Cohen, J. 1988. *Statistical Power Analysis for the Behavioral Sciences.* Hillsdale, NJ: Lawrence Erlbaum.

Dornheim, D. 2008. *Prädiktion von Rechenleistung und Rechenschwäche: Der Beitrag von Zahlen-Vorwissen und allgemein-kognitiven Fähigkeiten.* Berlin: Logos.

Duncan, G. J., C. J. Dowsett, A. Claessens, K. Magnuson, A. C. Huston, P. Klebanov, L. S. Pagani, et al. 2007. "School Readiness and Later Achievement." *Developmental Psychology* 43 (6): 1428–1446.

Engel, M., A. Claessens, T. Watts, and G. Farkas. 2016. "Mathematics Content Coverage and Student Learning in Kindergarten." *Educational Researcher* 45 (5): 293–300.

Field, A. 2009. *Discovering Statistics Using SPSS.* Los Angeles, CA: Sage.

Garrote, A., E. Moser Opitz, and C. Ratz. 2015. "Mathematische Kompetenzen von Schülerinnen und Schülern mit dem Förderschwerpunkt geistige Entwicklung. Eine Querschnittstudie." *Empirische Sonderpädagogik* 7 (1): 24–40.

Gasteiger, H. 2015. "Early Mathematics in Play Situations: Continuity of Learning." In *Mathematics and Transition to School: International Perspectives*, edited by B. Perry, A. Gervasoni, and A. MacDonald, 255–272. Singapore: Springer.

Gasteiger, H., A. Obersteiner, and K. Reiss. 2015. "Formal and Informal Learning Environments: Using Games to Support Early Numeracy." In *Describing and Studying Domain-Specific Serious Games*, edited by J. Torbeyns, E. Lehtinen, and J. Elen, 231–250. Cham: Springer.

Gross, C., and H. G. Rossbach. 2011. "Frühpädagogik." In *Empirische Bildungsforschung. Gegenstandsbereiche*, edited by H. Reinders, H. Ditton, C. Gräsel, and B. Gniewosz, 75–86. Wiesbaden: VS Verlag für Sozialwissenschaften.

Grüssing, M., and A. Peter-Koop. 2008. "Effekte vorschulischer mathematischer Förderung am Ende des ersten Schuljahres: Erste Befunde einer Längsschnittstudie." *Zeitschrift für Grundschulforschung* 1 (1): 65–81.

Hauser, B. 2005. "Das Spiel als Lernmodus: Unter Druck von Verschulung - im Lichte der neueren Forschung." In *Bildung 4- bis 8-jähriger Kinder*, edited by T. Guldimann, and B. Hauser, 143–168. Münster: Waxmann.

Hauser, B., E. Rathgeb-Schnierer, R. Stebler, and F. Vogt, eds. 2015. *Mehr ist mehr. Mathematische Frühförderung mit Regelspielen.* Seelze: Klett/Kallmayer.

Hauser, B., F. Vogt, R. Stebler, and K. Rechsteiner. 2014. "Förderung früher mathematischer Kompetenzen." *Frühe Bildung* 3 (3): 139–145.

Jörns, C., K. Schuchardt, D. Grube, and C. Mähler. 2014. "Spielorientierte Förderung numerischer Kompetenzen im Vorschulalter und deren Eignung zur Prävention von Rechenschwierigkeiten." *Empirische Sonderpädagogik* 2014 (3): 243–259.

Kamii, C., and Y. Kato. 2005. "Fostering the Development of Logico-Mathematical Thinking in a Card Game at Ages 5–6." *Early Education and Development* 16 (3): 367–384. doi:10.1207/s15566935eed1603_4.

Krajewski, K. 2003. *Vorhersage von Rechenschwäche in der Grundschule.* Hamburg: Kovac.

Krajewski, K., G. Nieding, and W. Schneider. 2007. *Mengen, zählen, Zahlen: Die Welt der Mathematik verstehen (MZZ).* Berlin: Cornelsen.

Krajewski, K., G. Nieding, and W. Schneider. 2008. "Kurz- und langfristige Effekte mathematischer Frühförderung im Kindergarten durch das Programm 'Mengen, zählen, Zahlen'." *Zeitschrift für Entwicklungspsychologie und Pädagogische Psychologie* 40 (3): 135–146.

Krajewski, K., and W. Schneider. 2009. "Exploring the Impact of Phonological Awareness, Visual–Spatial Working Memory, and Preschool Quantity–Number Competencies on Mathematics Achievement in Elementary School: Findings from a 3-Year Longitudinal Study." *Journal of Experimental Child Psychology* 103 (4): 516–531.

Kruse, J. 2015. *Qualitative Interviewforschung.* Weinheim: Juventa.

Kuckartz, U. 2010. *Einführung in die computergestützte Analyse qualitativer Daten.* 3rd ed. Wiesbaden: VS Verlag für Sozialwissenschaften.

Lee, J. S., and H. P. Ginsburg. 2009. "Early Childhood Teachers' Misconceptions about Mathematics Education for Young Children in the United States." *Australasian Journal of Early Childhood* 34 (4): 37–45.

Link, M., F. Vogt, and B. Hauser. 2017. "Überzeugungen von Kindergartenlehrpersonen zur mathematischen Förderung im Kindergarten: Schweiz, Deutschland und Österreich im Vergleich." *Beiträge zur Lehrerbildung* 35 (3): 440–458.

McCray, J. S., and J.-Q. Chen. 2012. "Pedagogical Content Knowledge for Preschool Mathematics: Construct Validity of a New Teacher Interview." *Journal of Research in Childhood Education* 26 (3): 291–307. doi:10.1080/02568543.2012.685123.

Moser, U., and N. Bayer. 2010. *4 bis 8. Schlussbericht der summativen Evaluation. Lernfortschritte vom Eintritt in die Eingangsstufe bis zum Ende der 3. Klasse der Primarschule.* Bern: Schulverlag plus.

Moser, U., and S. Berweger. 2007. *Wortgewandt & zahlenstark. Lern- und Entwicklungsstand bei 4- bis 6-Jährigen.* St. Gallen: interkantonale Lehrmittelzentrale.

Moser Opitz, E. 2001. *Zählen, Zahlbegriff, Rechnen. Theoretische Grundlagen und eine empirische Untersuchung zum mathematischen Erstunterricht in Sonderklassen.* Bern: Paul Haupt.

Oppermann, E., Y. Anders, and A. Hachfeld. 2016. "The Influence of Preschool Teachers' Content Knowledge and Mathematical Ability Beliefs on their Sensitivity to Mathematics in Children's Play." *Teaching and Teacher Education* 58: 174–184.

Ramani, G. B., and R. S. Siegler. 2008. "Promoting Broad and Stable Improvements in Low-Income Children's Numerical Knowledge Through Playing Number Board Games." *Child Development* 79 (2): 375–394.

Sarama, J., and D. H. Clements. 2009. *Early Childhood Mathematics Education Research: Learning Trajectories for Young Children.* New York: Routledge.

Schuler, S. 2008. "Was können Mathematikmaterialien im Kindergarten leisten? - Kriterien für eine gezielte Bewertung." In *Beiträge zum Mathematikunterricht 2008*, edited by Eva Vásárhelyi. Hildesheim: Franzbecker.

Schuler, S. 2013. *Mathematische Bildung im Kindergarten in formal offenen Situationen - eine Untersuchung am Beispiel von Spielen zum Erwerb des Zahlbegriffs.* Münster: Waxmann.

Singer, E. 2013. "Play and Playfulness, Basic Features of Early Childhood Education." *European Early Childhood Education Research Journal* 21 (2): 172–184.

Sonnenschein, S., and C. Galindo. 2015. "Race/Ethnicity and Early Mathematics Skills: Relations between Home, Classroom, and Mathematics Achievement." *Journal of Educational Research* 108 (4): 261–277.

Stebler, R., F. Vogt, I. Wolf, B. Hauser, and K. Rechsteiner. 2013. "Play-Based Mathematics in Kindergarten. A Video Analysis of Children's Mathematical Behaviour While Playing a Board Game in Small Groups." *Journal für Mathematik Didaktik* 34 (2): 149–175.

Thiel, O. 2010. "Teachers' Attitudes Towards Mathematics in Early Childhood Education." *European Early Childhood Education Research Journal* 18 (1): 105–115.

Van den Heuvel-Panhuizen, M. 1995. "Leistungsmessung im aktiv-entdeckenden Mathematikunterricht." In *Am Rande der Schrift. Zwischen Sprachenvielfalt und Analphabetismus*, edited by H. Brügelmann, H. Balhorn, and I. Füssenich, 87–107. Lengwil am Bodensee: Libelle.

van Oers, B. 2010. "Emergent Mathematical Thinking in the Context of Play." *Educational Studies in Mathematics* 74 (1): 23–37.

Vogt, F., and K. Rechsteiner. 2015. "Regelspiele entwickeln." In *Mehr ist mehr. Mathematische Frühförderung mit Regelspielen*, edited by B. Hauser, E. Rathgeb-Schnierer, R. Stebler, and F. Vogt, 46–55. Seelze: Klett/Kallmayer.

Vu, J. A., M. Han, and M. J. Buell. 2015. "The Effects of In-Service Training on Teachers' Beliefs and Practices in Children's Play." *European Early Childhood Education Research Journal* 23 (4): 444–460.

Weisberg, D. S., K. Hirsh-Pasek, and R. M. Golinkoff. 2013. "Guided Play: Where Curricular Goals Meet a Playful Pedagogy." *Mind, Brain, and Education* 7 (2): 104–112.

Weisberg, D. S., A. K. Kittredge, K. Hirsh-Pasek, R. M. Golinkoff, and D. Klahr. 2015. "Making Play Work for Education." *Phi Delta Kappan* 96 (8): 8–13.

Weiss, R. H., R. B. Cattell, and J. Osterland. 1997. *CFT 1. Grundintelligenztest Skala 1*. Göttingen: Hogrefe.

Wood, E. 2009. "Conceptualising a Pedagogy of Play: International Perspectives from Theory, Policy and Practice." In *From Children to Red Hatters: Divers Images and Issues of Play*, edited by D. Kuschner, 166–190. Lanham: University Press of America Inc.

Wullschleger, A., and R. Stebler. 2016. "Individuelle mathematikbezogene Lernunterstützung bei Regelspielen zur Förderung früher Mengen-Zahlen-Kompetenzen im Kindergarten." In *Perspektiven mathematischer Bildung im Übergang vom Kindergarten zur Grundschule*, edited by S. Schuler, C. Streit, and G. Wittmann, 171–186. Berlin: Springer Spektrum.

10 Using a bioecological framework to investigate an early childhood mathematics education intervention

Bob Perry and Sue Dockett

ABSTRACT

Over the last 20 years, the authors have utilised Bronfenbrenner's ecological and bioecological models as a basis for their work investigating children's transition to school, including the place of mathematics learning in this transition. The later bioecological model gave increased emphasis to the role of the individual within contexts, the processes that characterised interactions within and across contexts (proximal processes), and the influence of time. This bioecological model outlined four elements – person, process, context and time – which, together, were described as influencing the development of individuals. While the mathematical learning of young children influences, and is influenced by, all four elements of the model, the critical role of proximal processes in this learning is highlighted in this paper. Our aim is to identify how the four elements of the bioecological model, particularly proximal processes, provide a framework to analyse the experiences of the adults – early childhood educators and parents – involved in an early childhood mathematics education intervention designed to promote engagement with mathematics in playful situations. Data are drawn from 35 early childhood educators and 37 parents over 2 consecutive years (2013, 2014) with generally different participants in each year.

Introduction

This paper uses Bronfenbrenner's bioecological model of human development (Bronfenbrenner and Morris 2006) to analyse the results of an evaluation of a preschool mathematics education intervention in low socio-economic communities in Australia. The intervention was designed to enhance interactions in families among preschool-age (3–5 years old) children and adults in order to build children's positive dispositions to mathematics as they approached the start of school. We begin the paper with an introduction to Bronfenbrenner's bioecological model, noting its genesis through his earlier ecological model (Bronfenbrenner 1979). We then consider the application of the bioecological model to studies of transition to primary school, thus placing them within our extensive work in this field, and situating the study reported here within our ongoing work. The preschool mathematics education intervention and some results from the evaluation are then presented with specific consideration of the importance of the key constructs from the bioecological model.

Ecological and bioecological models

With the focus of our initial work in transition to school being to seek multiple perspectives and to encourage consideration of the contexts involved (Dockett and Perry 2001), we were drawn initially to Bronfenbrenner's (1979) ecological model of human development and, in later work, to his bioecological model (Bronfenbrenner and Morris 2006). The bioecological model builds upon, extends and refines the earlier ecological model. The changing nature of the models, and of our use of them (Dockett and Perry 2001, 2007, 2014a, 2014b), is a valuable reminder that theories can be dynamic, rather than static, as they are tried, tested, refined, applied in different ways and reformulated over time (Einarsdóttir 2014).

Ecological model

While described as a theory of human development, Bronfenbrenner's ecological theory (1979) attended not only to the individual – located at the centre of his concentric systems diagram – but also to the influence of the environments in which the individual was located. Bronfenbrenner's ecological model emphasised the importance of understanding individuals within their (multiple) environments. By drawing on the term 'ecology' to describe his model, Bronfenbrenner emphasised the interactions between individuals and their environment as key contributors to development. Further details concerning the ecological model are available from many sources (Bronfenbrenner 1979, 1994; Dunlop 2014; Rogoff 2003).

Critiques of the ecological model contested the representation of the individual at the centre of multiple contexts, arguing that not all contexts prioritised the individual, and that the model did not give adequate consideration to social and cultural constructs, or to power relations within contexts (Petriwskyj 2014; Rogoff 2003). Further, Vogler, Crivello, and Woodhead (2008, 25) argued that 'while the identification of multiple interacting systems is conceptually elegant, there is a risk of objectifying boundaries and assuming internal sub-system coherence'. In other words, we should not be surprised when boundaries between systems are blurred, or expect that microsystems operate in similar ways for all individuals.

Refinements to the ecological model (Bronfenbrenner 1988, 1994) added increased emphasis to the role of the individual within contexts, the processes that characterised interactions within and across contexts (proximal processes) and the influence of time. Rosa and Tudge (2013) note that these emphases reflected theoretical and conceptual elaborations derived from Vygotsky's (1962, 1978) notions of cultural contexts and the dialectic between individual and the environment; Lewin's (1936) attention to the concept of life space; Elder's (1998) life course theory; and Ceci's (Bronfenbrenner and Ceci 1994) formulation of the importance of proximal processes.

Bronfenbrenner's early model (1979) identified ecological transitions as normative changes that occurred within people's lives, and that required some form of adaptation on the part of the individual and/or the environment. Transitions were described as happening primarily at the level of the mesosystem as individuals interacted in different contexts or ecologies. His argument was underpinned by three characteristics of ecological environments: (i) the interdependence of systems, whereby what occurred – or did not occur – in one context was influenced by what happened in other contexts; (ii) the importance of interactive processes between and among people in facilitating change and (iii) each of those involved in the setting understands and perceives actions and interactions in a personal and unique way. As a consequence, it was argued that the understandings of individuals contribute to the setting and their perceptions of it.

Bioecological model and transition to school

The bioecological model (Bronfenbrenner and Morris 2006) outlined four elements – person, process, context and time (PPCT) – which, together, were described as influencing the development of individuals. We explore each of these elements below and consider the ways in which we have used them in our transition to school work, situating the mathematics education intervention in time, as children prepare to start school; in the contexts of families and educational settings; and through interactions between children and adults and among adults as the children start school.

Person characteristics of the individual influence developmental outcomes. In any situation, individuals bring with them a range of personal characteristics drawn from their biological as well as their experiential history. They include characteristics in the categories of demand, resource and force (Bronfenbrenner and Morris 2006). Demand characteristics – such as temperament, age, gender and appearance –may influence not only the ways in which individuals engage in interactions, but also the ways in which others interact with them. In considering starting school, a child's age or gender, for example, can influence educators' interactions with and expectations of them (Dockett and Perry 2002; Graue 1993).

Resource characteristics include 'the mental and emotional resources such as past experiences, skills, and intelligence and also the social and material resources' of individuals (Tudge et al. 2009, 200). When considering the transition to school, some of our colleagues have utilised the term 'virtual backpacks' to encompass resource characteristics (Margetts 2003; Peters 2014; Peters et al. 2009).

Force characteristics relate to dispositions, influencing each individual's motivation, persistence, curiosity and the like (Bronfenbrenner and Morris 2006). Factors such as children's responsiveness to others – adults and children – and the ease with which they form positive connections with others in new contexts are likely to influence their transitions to school. As well, the willingness to take risks and persist in learning tasks are included in this category.

Proximal processes are defined as the

> progressively more complex reciprocal interactions between an active, evolving, biopsychological human organism and the persons, objects, and symbols in its immediate external environment. To be effective, the interaction must occur on a fairly regular basis over extended periods of time. (Bronfenbrenner and Morris 2006, 797)

Key characteristics of proximal processes are their increasing complexity, reciprocal nature, interactive basis and regularity (Jaeger 2016). Proximal processes occur within relationships – not only with people, but also with objects and symbols. In relation to transition to school, proximal processes could include the interactions between parents and children as they 'prepare' for school; conversations and interactions with other children about what school is, or might be, like; and pedagogical strategies employed by educators in different settings. Proximal processes play an important role in helping individuals 'come to understand their world and formulate ideas about their place within it' (Tudge et al. 2009, 200). The impact of proximal processes is dependent on the other three elements identified – the characteristics of the developing person; the environments in which the actions and interactions occur; and their timing (Bronfenbrenner and Morris 1998). As well, the effectiveness of proximal processes can

> depend, to a substantial degree, on the availability and involvement of another adult, a third party, who assists, encourages, spells off, gives status to, and expresses admiration and affection for the person caring for and engaging in joint activity with the child. (Bronfenbrenner and Morris 2006, 823)

Such 'third parties' – early childhood educators, school teachers and parents – are very important in a child's transition to school.

Context was a predominant feature of ecological theory, with its attention to micro-, meso-, exo- and macrosystems. The importance of systems (contexts) in bioecological theory remains, with microsystems identified as primary sites for proximal processes. Despite this, what occurs within one system can influence what occurs within other systems, and experiences from several systems can generate both consistency and tension. For example, experiences within the mesosystem created when the microsystems of school, prior-to-school and home overlap, can be particularly important in supporting children and families as they manage the transition to school (Dockett and Perry 2007).

Time was included as one of the core elements of bioecological theory and was systematised into the theory through the introduction of the chronosystem (Bronfenbrenner 1988). The prominence of time was highlighted with references to microtime – the consistency of proximal processes; mesotime – how often proximal processes occur; and macrotime – historical time (Rosa and Tudge 2013). This focus emphasised both time and timing: not only what happened in the present, but also what had happened in the past and what was likely to happen in the future (Bronfenbrenner and Morris 2006). Consideration of time contributes to understanding issues of continuity and change. Just as individuals change, so too do contexts and cultures.

Each of the elements of PPCT is important within Bronfenbrenner's bioecological model. However, Bronfenbrenner came to view proximal processes as the driving factor in development (Jaeger 2016).

The bioecological model has underpinned much of our work in transition to school. We have noted previously that

> it prompts attention to the relationships and interactions associated with starting school, the characteristics and resources each individual (be they a child, family member, or educator) brings with them to the transition, recognition of the various systems or contexts in which children and families are located, as well as attention to specific events, patterns of interactions and historical context. (Dockett, Petriwskyj, and Perry 2014, 4)

Mathematics education plays a role in effective transition to school (Perry, MacDonald, and Gervasoni 2015). In the remainder of this paper, we focus on one preschool mathematics education initiative and the ways in which the PPCT model has provided the theoretical and analytical framework for the initiative.

The Smith Family and *Let's Count*

In 2010, The Smith Family[1] embarked on a project designed to promote young children's mathematics learning, based on evidence that, on average, children living in disadvantaged communities do not perform as well academically as children of the same age living in more advantaged communities (Carmichael, MacDonald, and McFarland-Piazza 2013; Caro 2009). It is also known that there is great potential for children living in low-income communities to benefit from mathematics intervention programmes (Sarama and Clements 2015). Evidence such as this, along with their significant social justice commitment, led The Smith Family to begin the development of the *Let's Count* programme focused on early years mathematics learning, with the aim of promoting children's positive dispositions to learning mathematics prior to their beginning school.

Development and implementation of the Let's Count *programme*

Let's Count is designed to assist family members, supported by early childhood educators, help their young children (aged 3–5 years) play with, investigate and learn powerful mathematical ideas, with the aim of developing positive dispositions to learning as well as mathematical knowledge and skills. *Let's Count* relies on educators supporting family members to provide opportunities for children to engage with the mathematics present in their everyday lives, talk about it, document it and extend it in ways that are relevant. The programme draws from bioecological theory (Bronfenbrenner and Morris 2006) and the importance of play in young children's mathematics learning (Siraj-Blatchford and Sylva 2004; Worthington and van Oers 2016). *Let's Count* was developed through The Smith Family by the first author of this paper and Ann Gervasoni.

Let's Count involves professional learning for early childhood educators that aims to enhance mathematics learning and teaching and strengthen partnerships between early years educators and parents by focussing on everyday opportunities for mathematics and opportunities for educators to consider how they might engage with parents to support children's mathematics learning. Ongoing interactions between educators, parents and children over the educational year follow from this professional learning. The programme encourages educators to use a variety of strategies to connect with families and stimulate mathematics learning. The key message in the *Let's Count* programme is *Notice, Explore and Talk about Mathematics*.

Evaluation of a pilot found that *Let's Count* assisted early childhood educators and parents to promote children mathematical engagement, learning outcomes and dispositions (Perry, Gervasoni, and Kearney 2012). *Let's Count* was refined and a further pilot programme was conducted during 2013 and 2014. Outcomes of the longitudinal evaluation of the programme have been reported extensively (Gervasoni and Perry 2016; Perry et al. 2016). This paper uses results of this evaluation to illustrate the importance of the bioecological model in understanding the impact of *Let's Count*.

Evaluation of Let's Count

The longitudinal evaluation of *Let's Count* used a multi-methods paradigm to answer the following three questions:

(1) How does participation in *Let's Count* impact on children's numeracy knowledge and dispositions as they make the transition to school?
(2) What is the impact of *Let's Count* on the educator participants' knowledge, interest and confidence in mathematics learning and teaching?
(3) What is the impact of *Let's Count* on families' confidence, and knowledge about noticing, investigating and discussing mathematics with their children?

Answers to these questions have been reported earlier (Gervasoni and Perry 2016; Gervasoni, Perry, and Parish 2015; The Smith Family 2015), along with the impact and future trajectory for *Let's Count* (Gervasoni and Perry 2016; Perry et al. 2016). In this paper, however, the characteristics of *Let's Count* and the data from the adults involved in the longitudinal evaluation are reported through the four elements of Bronfenbrenner's bioecological model, with the aspiration that this might stimulate other early childhood mathematics education researchers to consider utilising the model in their work.

Data from early childhood educators and parents were generated during telephone interviews about the implementation and impact of *Let's Count*. Educators and the families with whom they worked were interviewed twice in 2013, and the 2014 cohort was interviewed three times. In each year, the first interview occurred soon after the first professional learning workshop for educators (Gervasoni and Perry 2016). In total, 101 educator interviews and 125 parent interviews were conducted.

Let's Count and the bioecological model

Data from the interviews were coded under the four elements of the bioecological model. This proved a challenging assignment as many of the statements made by educators or parents displayed aspects of more than one element. Nonetheless, for clarity, each data statement will be presented under only one of the elements.

Person

Each person – child, parent, family member or educator – involved in *Let's Count* brings a range of person characteristics to their understandings of mathematics. These characteristics also 'invite or discourage reactions from the social environment' (Bronfenbrenner and Morris 1998, 1011). For example, adults' previous experiences with mathematics contribute to their person characteristics, including their motivation to engage with mathematics, as well as dispositions such as curiosity and persistence. Educators noted the change in their own attitudes towards mathematics:

> … I used to think of maths as sums. You know, when you think of maths you think of sums, like sitting at a high school desk trying to do these sums that you can't work out. But having now looked at maths in a different way I kind of see that it is everywhere and we do use it every day. So I'm starting to feel a bit more confident with that.

Parents, too, reflected on the impact of their prior experience on their confidence with mathematics:

> I would say I never was probably really good at maths at school. It took me forever to get fractions, but apart from that I was middle of the range in maths I suppose. Not my strength and not my weakness either. … probably at primary school level [I enjoyed maths]. Secondary it got a bit intense for my liking. But primary school was good.

Educators also reported changes in their attitudes towards children's capabilities to engage with mathematics:

> I'm really seeing the children … Just their knowledge has just blown me away, of what concepts they're understanding. Their understanding of like symmetry and patterns. And now it's starting to be more about adding. Last week we worked out that 10×3 is 30 and that was from a story book that was '10 Red Apples' from Dr Seuss. They had noticed that there was 10 on each of the animal's heads and then I chose three children to show me 10 [on their] hands. And then they were able to count along and find out that that actually meant 30 apples in total. And that all came from the children.

This same growing awareness of children's capabilities was echoed by parents:

> I think one of the things that surprised me lately is her ability … . She's been asking me to join numbers together to tell her what they equal. So she said 'If I've got 5 and 5 how many is that?' This is on the way home in the car from preschool actually, 'And 6 and 6 and how many is that?' And 'How many are my toes and how many are my fingers?'

While there is evidence from the interview data that all three categories of person characteristics – demand, resource and force – play a part in the success of *Let's Count*, demand characteristics such as age and gender seemed less important than matters of experience and disposition.

Process

In *Let's Count*, proximal processes are the central driver around the development of the programme, as relationships are built within groups of children, among educators and within families, as well as between members of each of these groups. *Let's Count* ran for about 9–10 months in each setting during the pilots in 2013 and 2014. This provided an extended period of time during which proximal processes could be established and enhanced.

One of the principles underpinning *Let's Count* is that, with appropriate support, family members are capable of noticing, supporting and challenging children's mathematical learning on a regular basis. A major influence on that family support comes from the educators, who themselves engage in proximal processes with family members, as well as the children. Hence, proximal processes are integral to *Let's Count* in three ways: in promoting parents/family members' confidence and understandings of young children's mathematics; supporting families to build and use a range of proximal processes in interactions with their children; and facilitating educators' increasingly complex interactions with children. Both parents and educators witnessed these processes.

Parents and educators both commented on the value of their interactions – many of which constitute proximal processes involving collaboration, feedback, and suggesting follow-up materials or resources. Often the educator and parents are playing the 'third party' role highlighted by Bronfenbrenner and Morris (2006):

> I think we're probably just getting a bit more engaged with the preschool because they are trying to involve us a bit more in the activities and I feel like there's a bit more ... Because of the [*Let's Count*] program and what they're trying to achieve with it, there's probably a few more things going on and they're probably giving me more feedback about what they're trying to teach the kids as a result of the program. (Parent)

> I like the way the parents are really involved and it's more about them, because that will hopefully continue on for the rest of their child's schooling; and for other children that they may have in their family as well. (Educator)

These interactions are possible because of the relationships that have been built or extended between parents/family members and educators.

> I think it's been a positive thing for building relationships with parents because they've felt that we're acknowledging them as their child's first educator for their own child. And you know, respecting the ideas that they have, like we're not pretending that we're the experts, we're asking them for their ideas and passing those on to other people and even using some of them here. So I think it's a good way to build positive relationships with parents. (Educator)

The experiences underpinning *Let's Count* were designed to engage strategies previously identified as effective in promoting proximal processes (Jaeger 2016), including drawing on children's everyday experiences – particularly play – to emphasise active involvement, both in the experiences and in the processes of meaning-making; engaging in authentic experiences using mathematics; encouraging children to pose questions and set goals that can be addressed through mathematics; highlighting the importance of communication and enjoyment; and providing feedback and encouragement. These elements are noted in the following example:

> [One child] wants to measure his bed, the information came from his mother first and then we discussed it with the child. The mother came in and said 'Oh he really wants to measure his bed' and I went 'OK, we can do that, we can work out a way to do that with you'. So ... it depends on developing a rapport between the educator and the parent through discussion. (Educator)

Context

Processes that support the development of mathematics occur within microsystems – such as the home, the preschool, the local community, as well as cultural and social groups. For the child, becoming a mathematician involves all of these contexts – even though experiences may occur only within one microsystem, becoming a mathematician 'occurs, in a conceptual sense, at the intersection of all the microsystems' that involve the child (Jaeger 2016, 178). Becoming a mathematician occurs within a mesosystem.

Illustrative of the significance of the mesosystem, parents noted occasions when children drew on experiences within the preschool microsystem to understand what was happening at home.

> It's been a lot of noticing things in her surroundings that I don't think she would have noticed before, based on the fact that she has had exposure to the words and the language and the concepts ... the other day she just noticed a clock they had at Bunnings [hardware store]

and then she was trying to tell me the time and talking about the hands and things like that. Even playing games she uses language like halves and that's a quarter. Before she would never have, never, you know, been talking like that. (Parent)

In similar ways, educators took note of children's experiences in other microsystems to inform their own actions:

Well a lot of the children play soccer games or a sports game where they see a team, a player wearing a number … We had lots of children who recognised numbers from where they've been in their life, like bus numbers. We added timetables to their play area so they could see numbers. And then we talked about time as well. Because that sort of all came around because of the catching a bus, buses come at a certain time. (Educator)

Among the early childhood educators in particular, there developed a belief that there was mathematics in everything, in every microsystem. Context was central to a number of the comments made by both parents and educators:

I suppose what we've taken away from going to the training the other day is that maths is in everything you do. It's just making it more visible. (Educator)

Nowadays with the family just baking a cake or just hanging out the washing to understand that you're actually encouraging your child to do mathematics and stuff like that well you know, it gives you more of a push to encourage your child to do it. (Parent)

Time

The element of time and its influence on children's developing mathematical understandings was noted by parents and educators. Reference to microtime was concerned primarily with the contingency of responses from adults – with both parents and educators describing the importance of seizing the moment and responding when children indicated an interest in mathematical learning, such as when a question was asked in the car or when a child sought specific involvement from an educator. There was also recognition that the building of relationships among people takes time:

One little boy this morning said to me 'Look what I've made, come and see, we've made a really long thing' and I said 'How long is it?' and he said 'Well it's longer than this'. We kept going on about it, I said 'What have you used?', he said 'I made some long and some short blocks' and I said 'Well what else can we do with it?'. (Educator)

References to macrotime were of two main types: reflections on what mathematics was like when adults were at school and comments relating to the current educational landscape, particularly strategies to have children 'ready for school', in the context of perceptions of high academic expectations for young children and the challenges adults associated with children 'falling behind':

I've got a son who can count to 100 and he asked me today if I could count to 100. I said 'Yes, I can count to 100 and I can even count to 1000' and he was just like so astonished, 'I'll never be able to count to 1000'. … He's 4, he's not 5 till the end of the year. … So just acknowledging to him too that one day he'll be able to do it, so that was quite cool. (Educator)

I think that will help them when they get to school, hopefully the maths terms that they're hearing aren't new to them because they've already heard them and know a little bit about what it might be about before they get there. So, it's not just all brand new stuff. (Educator)

Discussion

The elements of the bioecological model can be used to analyse the *Let's Count* programme and to consider the reasons for its success. However, it is the confluence of these elements that provides the greatest explanatory power in considering the impact of the programme. These elements are integrated into the mantra of *Let's Count*: *Notice, Explore and Talk about Mathematics.*

Notice

Many of the adult participants were surprised by the mathematical thinking of the young children with whom they interacted through *Let's Count*. The resource characteristics of the children were noticed by the adults and this noticing prompted them to provide opportunities for the children to continue their exploration of mathematical ideas. The adults were also noticing how the *Let's Count* programme was helping them to change their and the children's force characteristics, such as their attitudes towards mathematics, and their dispositions to be involved in the children's mathematics learning. Through their involvement as 'third parties', there also seemed to be a reconsideration of the importance of each of the adult groups – parents and educators – in each other's eyes.

One of the key messages of *Let's Count* is that there is mathematics in everything. On several occasions, educators and parents reported noticing mathematics in their own environments and also reported children doing the same. However, given the different *person* characteristics involved, there is often different mathematics noticed.

Noticing mathematics takes *time*, in both micro and macro forms. Many of the parents and educators have seen the value of *Let's Count* continuing into the future. They have also projected from past experiences and compared how they felt about mathematics then and after experiencing *Let's Count*.

Explore

Let's Count emphasises the role of play and investigation in children's learning. The programme is firmly based on the belief that young children learn through play, particularly with scaffolded support from more knowledgeable others (DEEWR 2009). Noticing more of what the children are thinking about means that adults are becoming aware of the 'invisible' exploration that children undertake.

When children or adults notice mathematics in their contexts, there is often strong motivation to continue exploring that mathematics. For example, rote counting, while perhaps not the most illuminating mathematical activity possible, does provide a challenge. Provided this challenge remains a personal one rather than a competition, for either children or parents, exploration can continue.

Talk about

It has long been known that the active use of language is important in the development of mathematical ideas (Ellerton and Clements 1991; Riccomini et al. 2015). This importance is emphasised in the Australian curriculum framework for the early years (DEEWR 2009) and was one of the major themes from both parents and educators in the evaluation of *Let's Count*. 'Talking about' mathematics in everyday lives encourages all participants to sustain powerful interactions over extended periods of time. Increases in confidence and capability among adults and children have helped develop the complexity of these interactions. Both parents and educators have remarked on the capabilities of young children to deal with such complexity. 'Talking about' mathematics has clearly been a key proximal process throughout *Let's Count*.

Role of proximal processes

In the bioecological model, Bronfenbrenner identified proximal processes as the 'engine' driving development. While each element of *Let's Count* contributes fuel to this engine, the processes of *notice, explore* and *talk about* afford many opportunities to engage proximal processes. Following Bronfenbrenner's lead, we argue that effective interactions to support the mathematical learning of young children involve proximal processes, built on a framework of noticing children, their interests, their existing understandings and individual characteristics, the contexts in which they are located and their situatedness in time. In other words, effective processes are proximal processes that take into account the person, context and time elements of the bioecological model.

Proximal processes involve children interacting, over time, with others and with a range of materials and resources – including mathematical symbols and concepts – located within specific contexts. Becoming a mathematician occurs within mesosystems, where, conceptually, children recognise and engage with mathematics that may have occurred in one context, but that has relevance and application in other contexts. Supporting children's mathematical development requires reciprocal interactions: reciprocal in the sense that all participants make an active contribution to the interaction; and in the sense that interactions are responsive, building on and extending children's understandings. Such responsiveness is facilitated when the adults in children's lives acknowledge their existing mathematical understandings and can call on appropriate resources to challenge and extend these.

The most effective proximal processes tend to occur between those who have strong relationships (Jaeger 2016). From this basis alone, it would seem essential to engage families in the promotion of young children's mathematical development. However, family members are not the only ones who may have strong relationships with young children. Educators and peers too can engage in sensitive and responsive interactions with children who enable them to recognise existing competencies and to extend the complexity of interactions over time. The opportunity for each of the adult groups to act as a 'third party' to the other in interactions with children has been critical to the success of *Let's Count*.

Conclusion

Bronfenbrenner's bioecological model did not claim to be a theory of mathematical development. However, the elements of the PPCT model can be used to explore what individuals bring to their mathematics learning, processes that underpin such learning, contexts in which such learning occurs, and the importance of both time and timing in developing understandings. These elements have been integrated into the *Let's Count* programme, through the mantra *Notice, Explore and Talk About Mathematics,* recognising that supporting children's mathematical development involves working collaboratively with those who are in a position to facilitate meaningful, ongoing, regular, reciprocal and increasingly complex interactions with mathematics at their core. While early childhood educators in contexts outside Australia may not be able to adopt the complete *Let's Count* programme, they can adopt the mantra in their own mathematics programmes, assured that it is closely tied to the elements of the bioecological model.

Note

1. The Smith Family is an Australian charity dedicated to supporting the education of children who live in communities facing multiple disadvantages.

Acknowledgements

The authors acknowledge the support of Blackrock Investment Management (Blackrock Australia 2017), the Origin Foundation (2017) and The Smith Family (2017) in the development and evaluation of *Let's Count*. Special acknowledgement is given to Ann Gervasoni for her work in developing and evaluating the programme.

Disclosure statement

No potential conflict of interest was reported by the authors.

References

Blackrock (Australia). 2017. "About us." Accessed January 22, 2017. https://www.blackrock.com/au/individual/about-blackrock.

Bronfenbrenner, U. 1979. *The Ecology of Human Development: Experiments in Nature and Design.* Cambridge, MA: Harvard University Press.

Bronfenbrenner, U. 1988. "Interacting Systems in Human Development: Research Paradigms Present and Future." In *Persons in Context: Developmental Processes*, edited by N. Bolger, A. Caspi, G. Downey, and M. Moorhouse, 25–49. Cambridge: Cambridge University Press.

Bronfenbrenner, U. 1994. "Ecological Models of Human Development." In *International Encyclopedia of Education*, edited by T. Postlethwaite and T. Husen, Vol. 3, 2nd ed., 1643–1647. Oxford: Elsevier.

Bronfenbrenner, U., and S. J. Ceci. 1994. "Nature-Nuture Reconceptualized in Developmental Perspective: A Bioecological Model." *Psychological Review* 101: 568–586.

Bronfenbrenner, U., and P. A. Morris. 1998. "The Ecology of Developmental Processes." In *Handbook of Child Psychology: Vol. 1. Theoretical Models of Human Development*, 5th ed., edited by W. Damon and R. M. Lerner, 993–1028. New York: John Wiley & Sons.

Bronfenbrenner, U., and P. A. Morris. 2006. "The Bioecological Model of Human Development." In *Handbook of Child Psychology, Vol. 1: Theoretical Models of Human Development*, 6th ed., edited by W. Damon and R. M. Lerner, 793–828. New York: Wiley.

Carmichael, C., A. MacDonald, and L. McFarland-Piazza. 2013. "Predictors of Numeracy Performance in National Testing Programs: Insights from the Longitudinal Study of Australian Children." *British Educational Research Journal.* doi:10.1002/berj.3104.

Caro, D. H. 2009. "Socio-Economic Status and Academic Achievement Trajectories from Childhood to Adolescence." *Canadian Journal of Education* 32 (3): 558–590.

DEEWR (Department of Education, Employment and Workplace Relations). 2009. *Belonging, Being and Becoming: The Early Years Learning Framework for Australia.* Canberra: Commonwealth of Australia. Accessed February 14, 2017. http://www.deewr.gov.au/earlychildhood/policy_agenda/quality/pages/earlyyearslearningframework.aspx.

Dockett, S., and B. Perry, eds. 2001. *Beginning School Together: Sharing Strengths.* Watson: Australian Early Childhood Association.

Dockett, S., and B. Perry. 2002. "Who's Ready for What? Young Children Starting School." *Contemporary Issues in Early Childhood* 3 (1): 67–89.

Dockett, S., and B. Perry. 2007. *Transitions to School: Perceptions, Experiences and Expectations.* Sydney: University of New South Wales Press.

Dockett, S., and B. Perry. 2014a. "Mapping Transitions." *Early Childhood Folio* 18 (2): 33–38.

Dockett, S., and B. Perry. 2014b. "Research to Policy: Transition to School Position Statement." In *Transitions to School – International Research, Policy and Practice*, edited by B. Perry, S. Dockett, and A. Petriwskyj, 277–294. Dordrecht: Springer.

Dockett, S., A. Petriwskyj, and B. Perry. 2014. "Theorising Transition: Shifts and Tensions." In *Transitions to School – International Research, Policy and Practice*, edited by B. Perry, S. Dockett, and A. Petriwskyj, 1–18. Dordrecht: Springer.

Dunlop, A.-W. 2014. "Thinking about Transitions: One Framework or Many? Populating the Theoretical Model over Time." In *Transitions to School – International Research, Policy and Practice*, edited by B. Perry, S. Dockett, and A. Petriwskyj, 31–46. Dordrecht: Springer.

Einarsdóttir, J. 2014. "Readings of Media Accounts of Transition to School in Iceland." In *Transitions to School – International Research, Policy and Practice*, edited by B. Perry, S. Dockett, and A. Petriwskyj, 21–30. Dordrecht: Springer.

ElderJr., G. H. 1998. "The Life Course as Developmental Theory." *Child Development* 69: 1–12.

Ellerton, N. F., and M. A. Clements. 1991. *Mathematics in Language: Language Factors in Mathematics Learning.* Geelong: Deakin University.

Gervasoni, A., and B. Perry. 2016. "The Impact on Learning When Families and Educators Act Together to Assist Young Children to Notice, Explore and Discuss Mathematics." In *Mathematics Education in the Early Years*, edited by T. Meaney, O. Helenius, M. L. Johansson, T. Lange, and A. Wernberg, 115–135. Dordrecht: Springer.

Gervasoni, A., B. Perry, and L. Parish. 2015. "The Impact of *Let's Count* on Children's Mathematics Learning." In *Mathematics Education in the Margins* (Proceedings of the 38th Annual Conference of the Mathematics Education Research Group of Australasia), edited by M. Marshman, V. Geiger, and A. Bennison, 253–260. Sunshine Coast: MERGA.

Grauc, M. E. 1993. *Ready for What? Constructing Meanings of Readiness for Kindergarten.* Albany: State University of New York Press.

Jaeger, E. L. 2016. "Negotiating Complexity: A Bioecological Systems Perspective on Literacy Development." *Human Development* 59: 163–187.

Lewin, K. 1936. *Principles of Topological Psychology.* New York: McGraw-Hill.

Margetts, K. 2003. "Children Bring More to School than Their Backpacks: Starting School Down Under." *Journal of European Early Childhood Education Research Monograph* 1: 5–14.

Origin Foundation. 2017. "Who We Are." Accessed April 21, 2017. http://originfoundation.com.au/who-we-are.

Perry, B., A. Gervasoni, A. Hampshire, and W. O'Neill. 2016. "*Let's Count*: Improving Community Approaches to Early Years Mathematics Learning, Teaching and Dispositions Through Noticing, Exploring and Talking about Mathematics." In *Opening up Mathematics Education Research* (Proceedings of the 39th Annual Conference of the Mathematics Education Research Group of Australasia), edited by B. White, M. Chinnappan, and S. Trenholm, 75–84. Adelaide: MERGA.

Perry, B., A. Gervasoni, and E. Kearney. 2012. *Let's Count Pilot Program: Final Evaluation.* Sydney: The Smith Family. Unpublished Report.

Perry, B., A. MacDonald, and A. Gervasoni, eds. 2015. *Mathematics and Transition to School – International Perspectives.* Dordrecht: Springer.

Peters, S. 2014. "Chasms, Bridges and Borderlands: A Transitions Research 'Across The Border' from Early Childhood Education to School in New Zealand." In *Transitions to School – International Research, Policy and Practice*, edited by B. Perry, S. Dockett, and A. Petriwskyj, 105–116. Dordrecht: Springer.

Peters, S., C. Hartley, P. Rogers, J. Smith, and M. Carr. 2009. "Early Childhood Portfolios as a Tool for Enhancing Learning during the Transition to School." *International Journal of Transitions in Childhood* 3: 4–15.

Petriwskyj, A. 2014. "Critical Theory and Inclusive Transitions to School." In *Transitions to School – International Research, Policy and Practice*, edited by B. Perry, S. Dockett, and A. Petriwskyj, 201–215. Dordrecht: Springer.

Riccomini, P. J., G. W. Smith, E. M. Hughes, and K. M. Fries. 2015. "The Language of Mathematics: The Importance of Teaching and Learning Mathematical Vocabulary." *Reading and Writing Quarterly* 31 (3): 235–252.

Rogoff, B. 2003. *The Cultural Nature of Human Development*. Oxford: Oxford University Press.

Rosa, E. M., and J. Tudge. 2013. "Urie Bronfenbrenner's Theory of Human Development: Its Evolution from Ecology to Bioecology." *Journal of Family Theory & Review* 5 (4): 243–258.

Sarama, J., and D. H. Clements. 2015. "Scaling up Early Childhood Mathematics Interventions: Transitioning with Trajectories and Technologies." In *Mathematics and Transition to School*, edited by B. Perry, A. MacDonald, and A. Gervasoni, 153–169. Dordrecht: Springer.

Siraj-Blatchford, I., and K. Sylva. 2004. "Researching Pedagogy in English Pre-schools." *British Educational Research Journal* 30 (5): 713–730.

The Smith Family. 2015. "Strengthening Early Numeracy Learning: The *Let's Count* Program." Accessed August 9, 2017. https://www.thesmithfamily.com.au/~/media/files/research-advocacy/research/lets-count-research.ashx.

The Smith Family. 2017. "About Us." Accessed October 6, 2017. https://www.thesmithfamily.com.au/about-us.

Tudge, J., I. Mokrova, B. Hatfield, and R. Karnik. 2009. "Uses and Misuses of Bronfenbrenner's Bioecological Theory of Human Development." *Journal of Family Theory and Review* 1: 198–210.

Vogler, P., G. Crivello, and M. Woodhead. 2008. *Early Childhood Transitions Research: A Review of Concepts, Theory and Practice*. The Hague: Van Leer Foundation.

Vygotsky, L. S. 1962. *Thought and Language*. Cambridge, MA: MIT Press.

Vygotsky, L. S. 1978. *Mind in Society: The Development of Higher Psychological Processes*. Cambridge, MA: Harvard University Press.

Worthington, M., and B. van Oers. 2016. "Pretend Play and the Cultural Foundations of Mathematics." *European Early Childhood Education Research Journal* 24 (1): 51–66.

Index